EUREKA

ALSO BY CHAD ORZEL

HOW TO TEACH RELATIVITY TO YOUR DOG
HOW TO TEACH PHYSICS TO YOUR DOG

EUREKA

DISCOVERING YOUR
INNER SCIENTIST

CHAD ORZEL

BASIC BOOKS
A MEMBER OF THE PERSEUS BOOKS GROUP
NEW YORK

Copyright © 2014 by Chad Orzel
Published by Basic Books,
A Member of the Perseus Books Group

Books published by Basic Books are available at special discounts for bulk
purchases in the United States by corporations, institutions, and other
organizations. For more information, please contact the Special Markets
Department at the Perseus Books Group, 2300 Chestnut Street, Suite 200,
Philadelphia, PA 19103, or call (800) 810-4145, ext. 5000, or e-mail special.
markets@perseusbooks.com.

Designed by Jack Lenzo

A CIP catalog record for this book is available from the Library of
Congress.

ISBN: 978-0-465-07496-9 (paperback)
ISBN: 978-0-465-04491-7 (e-book)
LCCN: 2014034615

10 9 8 7 6 5 4 3 2 1

For my parents,
whose unfailing support
made all of this possible.

CONTENTS

DISCOVERING YOUR INNER SCIENTIST

When I brought home my iPad for the first time, before I'd finished getting it out of the box, my then three-year-old daughter spotted it and announced, "I want to play Angry Birds!" I was a little bemused by the idea that the game's marketing reached even the preschool set, but I was not particularly unhappy. After all, Angry Birds is a great way to learn about science.

I'm not talking about using the game to investigate the physics of birds flung from slingshots, but I mean the *process* of the game: To succeed at Angry Birds, you need to think like a scientist.* When confronted with a new level, you need to look closely at the arrangement of pigs and blocks and other elements, to determine exactly what you need to accomplish. Then you develop a mental model of what will happen when you start launching birds: "If I hit this block with the yellow bird, it will topple that block onto the pigs, and collapse this tower. . . ." Then you test your model directly and see how

---

* A fellow physics professor and blogger does study bird-flinging physics and has even written a book about the physics of Angry Birds: Rhett Allain, *Angry Birds: Furious Forces* (Washington, DC: National Geographic, 2013).

well your prediction matches (video-game) reality. If you've guessed correctly, you devastate the pigs and move on to the next level; if your model was incorrect or incomplete, you refine it and try again.

That's as nice an encapsulation of the process of science as you'll find anywhere, and it's wildly popular. The original mobile-phone game and its many spin-offs are among the most successful video-game franchises of all time, downloaded over a billion times as of early 2013 and boasting more than 260 million monthly users. It's simple enough that a three-year-old can figure out the basics, and complex enough to be addictive even for adults. I've lost hours to it myself.

The success of Angry Birds gives the lie to the popular notion that science is something arcane and esoteric, beyond the comprehension of ordinary humans. Even people who openly state that they can't understand and don't like science play Angry Birds and, in the process, use exactly the same bag of mental tricks that scientists do as they try to figure out the basic rules governing the universe. One in every twenty humans alive today devote part of their idle time each month to acting like a scientist, just in the pursuit of one silly video game. All of us, from three-year-olds on up through adults, have a scientist inside us.

LOOK, THINK, TEST, TELL

That may seem like a strange thing to claim. When I say that everyone has an inner scientist, I get a lot of skeptical reactions from people who say, "I never use any of the stuff I learned in science class." This reaction confuses two different meanings of *science*: One is the *product* of science, the body of knowledge about particles, molecules, animals, planets, and stars—facts that people memorize in science classes—and the other is the *process* of science. The process is, to my mind, the more fundamental and important definition of science.

The process of science consists of four steps:*

- *Look* at the world around you, and identify some phenomenon you would like to understand.
- *Think* of a model that might explain how and why that phenomenon occurs in terms of general rules about how the universe operates.
- *Test* your theory with further observations, and carry out experiments to see if the predictions of your model agree with reality.
- *Tell* everybody you know your proposed explanation and the results of your tests.

Clearing a level on Angry Birds requires you to go through the steps of the scientific process.† You *look* at the level, *think* about a possible tactic for wiping out the pigs, and *test* your model by launching the birds at your selected target. And if you're playing with friends or kids, you *tell* them how you did

---

* This definition is far too vague to pass muster with a philosopher of science, of course, but I'm trying to be inclusive rather than exclusive for this book. I'm not concerned with the precise location of the boundary between science and not-science, but I want to capture the broad outline of the process. The full story is far more complicated than this simple description, but this basic outline is enough for the purposes of this book. For an excellent, more detailed look at how science is done, see the Understanding Science project at UC Berkeley (http://undsci.berkeley.edu /index.php).

† Unless, of course, you're just mechanically following the level-clearing instructions presented on one of the many cheat sites on the Web. But that's not very much fun.

it and share your results with other people via high-score lists and other social features.

This four-step process is the best method we have for generating reliable knowledge about the world. It's also one of the most fundamentally human activities there is, and this process and its products, for good or ill, have made us the dominant species on the planet.

SCIENCE IS WHAT MAKES US HUMAN

In 2011, scientists excavating the Blombos Cave in South Africa unearthed what they described as "prehistoric paint kits" dated to a hundred thousand years ago: shell containers and simple stone tools bearing traces of colored minerals. The traces on the tools show that these minerals were ground to powder, and traces in the containers indicate that the powder was mixed with charcoal and animal fat to make a paste. Previous excavations had discovered chunks of red ochre from about seventy-five thousand years ago with an angled pattern etched into one of the blocks. These are some of the earliest direct evidence we have of symbolic thinking in humans, with the production of decorative pigments and the deliberate, abstract marking of the raw materials.

These paint-making tools also offer clear evidence of prehistoric science. The minerals used to make the pigments were dug up several kilometers away and brought to the cave. The paste-making involved several components, which were processed separately, then deliberately combined. The traces on one of the grinding stones show that it was first used to grind bits of yellow goethite, then cleaned or resharpened by chipping stone flakes off the end, before being used to grind red ochre.

Taken all together, the artifacts show that as early as one hundred thousand years ago, humans were doing science.

They had developed, through trial and error, a complicated, multistep process for producing colored pastes. They had determined which minerals were needed to produce particular colors, located good sources of them, and worked out how to grind and mix them with other materials to produce the desired results. They even knew to use different types of minerals for different colors and took care to avoid mixing them. And they shared that information with others, passing it down through the millennia that the cave was occupied.

For as long as there have been humans, then, there have been humans doing science. The process of looking at the world, figuring out how things work, testing that knowledge, and sharing it with others ought to be taken as one of the defining traits of our species. The process of science is not some incidental offshoot of more general human activity; it's the very thing that makes us who we are.

Everything we are as a species can be traced back to the scientific process of discovery by trial and error. Stone tools and weapons let us become predators rather than prey, despite our not having much in the way of natural armament. Clothes made from animal skins and natural fibers let us expand into environments very different from the African plains where we evolved. The development of agriculture led to urban civilizations, which freed up some people to study the world more thoroughly, which fueled further advances, and so on until the modern day, when we thoroughly dominate the surface of the Earth and some distance above it. Not a bad end for some jumped-up apes playing with paint in seaside caves.

SCIENTISTS ARE NOT THAT SMART

The claim that science is fundamentally human may seem outlandish, given that very few people spend their days doing things that we recognize as science. The popular image of scientists is of a tiny, elite (and possibly deranged) minority of people engaged in esoteric pursuits. One of the three most common responses when I tell somebody I'm a physicist is, "You must be really smart. I could never do that."* Other scientists I've spoken with report the same.

While the idea that scientists are uniquely smart and capable is flattering to the vanity of nerds like me, it's a compliment with an edge. There's a distancing effect to being called "really smart" in this sense—it sets scientists off as people who think in a way that's qualitatively different from "normal" people. We're set off even from other highly educated academics—my faculty colleagues in arts, literature, and social science don't hear that same "You must be really smart" despite the fact that they've generally spent at least as much time acquiring academic credentials as I have. The sort of scholarship they do is seen as just an extension of normal activities, whereas science is seen as alien and incomprehensible.

A bigger problem with this awkward compliment, though, is that it's just not true. Scientists are not *that* smart—we don't think in a wholly different manner than ordinary people do. What makes a professional scientist is not a supercharged brain with more processing power, but a collection of subtle differences in skills and inclinations. We're slightly better at doing the sort of things that professional scientists do on a daily basis—I'm better with math than the average person—but more importantly, we enjoy those activities and so spend time honing those skills, making the differences appear even greater.

---

* The other responses are, "I hated that when I took it in high school/ college," and, "Can you explain string theory to me?" This goes a long way toward explaining why physicists have a reputation as lousy conversationalists.

To turn things around a bit, I'm a decidedly mediocre carpenter. Not because I lack any of the physical or mental skills needed for the task—I can and have built things out of wood for home improvement projects. I'm not good at it, though, because I don't particularly enjoy the process and so don't seek out opportunities to engage in carpentry. I can do it if I have to, but my work is slow and plodding, and when I try to speed it up, I make mistakes and end up needing to start over. Professional woodworkers or even serious hobbyists, who do enjoy those tasks and put in the time practicing them, are vastly better at the essential tasks. This is not merely physical, either—they're also better at the mental aspect of the job, figuring out how to accomplish a particular construction goal, which is where I generally fail most dramatically. But you'll never hear anyone say, "A carpenter? You must be really smart."

Inclination isn't the entire story, to be sure—innate talents play a role. I love the game of basketball and spend a good deal of time playing in a lunchtime pickup game with other faculty, staff, and students. No amount of practice is going to make me an NBA player, though, because I lack the necessary physical gifts—I'm considerably taller than average (about six feet six), but not especially quick or agile. I struggle to keep up with the players on the team at Union College and wouldn't stand a chance against a big-time college player. Love of the game only takes you so far.

At the same time, though, my lack of elite physical ability doesn't preclude my playing basketball for several hours a week. And nobody thinks it all that odd for a bunch of middle-aged professors to devote some time to running around playing games; there are even some social advantages to participating in sports on an amateur level, particularly for men. Nobody expects us to say, "Well, I can't make it to the NBA, so I'll never touch a basketball again."

And yet, that's just what happens with science, which is the biggest problem with the perception of scientists as really smart. The incorrect image of science as something only a few innately brilliant individuals can handle leads many people to give up on science entirely once they decide that they can't make a career of it. There's no social advantage to having an amateur interest in science; in some contexts, it's even a liability, marking you out as a nerd. This problem is unique to science—nobody thinks it odd for people who can't make a career in academia to continue to read literature, go to art museums, or take an active interest in politics and history. In many professional contexts, it's considered surprising if you *don't* do any of those things, but science is something you're expected to give up unless you're "really smart" enough to make it your career.

DISCOVERING YOUR INNER SCIENTIST

Ironically, though, even people who consciously reject the thought of doing science themselves spend a good deal of time acting like scientists. The scientific process is a major component of any number of popular hobbies and pastimes. Whatever you do to unwind, it almost certainly draws on the same bag of mental tricks used by successful scientists. If you collect stamps or coins as a hobby, you're making use of the same impulse that helped Charles Darwin develop the theory of evolution. If you play bridge or other card games, you use the same inference process that led Vera Rubin and other astronomers to the realization that the universe contains vast amounts of stuff we can't see directly. And if you do crossword puzzles to relax, you're using the cross-checking and deduction that led the founders of quantum mechanics to develop the strangest and most successful theory in the history of physics.

The rest of this book is broken into four parts, organized around the steps in the scientific process: looking, thinking, testing, and telling. Each chapter in a part will describe an everyday activity—antique hunting, cooking a meal, playing basketball—that relates to one of these aspects of scientific activity. We'll also look at historical examples of scientific discoveries that crucially relied on a process similar to that particular hobby.*

Through these stories, I hope to convince you that whether you realize it or not, you have an inner scientist. You're using the process of science every day, even if you don't do science for a living. In the same way that amateur athletes blow off steam by playing the occasional game of basketball, nonscientists unwind by using scientific thinking in the pursuit of fun and relaxation. Knowing this, I hope, will inspire you to make some more conscious use of your inner scientist, both to better understand science and to more effectively pursue other interests.

EUREKA

This book takes its title from one of the oldest and best-known anecdotes in the history of science, concerning the ancient Greek mathematician and engineer Archimedes of Syracuse. When Hiero II ascended to power in Syracuse around 265 BCE, he commissioned a wreath of gold to be offered as a tribute to the gods. After the wreath was completed, he became suspicious that the goldsmith had held back some of the metal

---

* Most of the historical examples will be drawn from physics and astronomy, because those are the fields I know best, but the overall process is universal, and analogous examples can be found in all fields of science.

and replaced it with an equal (but cheaper) weight of silver. Hiero asked Archimedes to find a way to determine whether the ornate wreath was, in fact, pure gold.

This is a tricky question, because the weight of the wreath matched the weight of gold the goldsmith had been given, and Archimedes worried over the problem for some time. One day after much fruitless thought, he settled into a hot bath and noticed that as he stepped into the tub, water spilled over the sides. And as he pushed more of his body into the tub, more water spilled over. This, he realized suddenly, was the solution to his problem: Since silver is less dense than gold, a wreath made of a mix of the two metals would have a greater volume than an equal mass of pure gold and would push aside more water when submerged. He leaped out of the tub yelling "Eureka!" (Greek for "I've found it!") and was so excited that he forgot to dress before racing to the royal palace to tell the king his solution.

This might seem like an awkward starting point for a book that intends to demystify the process. After all, as the story has it, Archimedes solves the problem via a flash of insight, something that seems awfully mystical and remote from everyday activities. Also, he's an unworldly weirdo who ends up running naked through the streets because he figured out a physics problem.

The "Eureka!" moment, though, is a much more general phenomenon, something that nearly everyone has experienced at some point. Indeed, Archimedes's exclamation has become a shorthand phrase for that moment when the solution to a vexing problem becomes clear, seemingly in an instant, or when a wholly new idea pops into mind. The creation of Barbie dolls, Starbucks shops, and any number of popular songs has been attributed to eureka moments while the discoverer was doing something seemingly unrelated to the problem at hand. For myself, a lot of my best writing ideas have come not while I was sitting at the keyboard staring at the computer screen, but while I was out walking the dog, or playing with my kids.

Like most things involving thinking about thinking, though, the feeling of out-of-nowhere inspiration is almost certainly an illusion, hiding a lot of behind-the-scenes mental activity. Archimedes didn't stumble across the solution to a problem he had no prior part in, after all. He noticed a small detail related to a problem he had been struggling with for some time, and that meshed with pieces already in place to provide the answer he needed. This wasn't a moment of isolated inspiration, but was the culmination of an ongoing process of thought. Archimedes didn't stop thinking like a scientist just because he was taking a bath, even if he was probably taking a bath as a much-needed break from worrying about a scientific problem.

Scientists can find inspiration in everyday activities largely because the thought processes involved in those activities are not all that far removed from those involved in doing science. And the reverse is true, as well: Many everyday activities can benefit from a scientific approach. Ideas for new toy lines or business plans might seem to come from a very distant place, but in reality, they're the product of our inner scientists working diligently in the background. And the systematic working out of those ideas to make massively successful and inescapable dolls and coffee shops has a lot in common with the process of science.

WHY SHOULD YOU THINK LIKE A SCIENTIST?

Even if everyone can think like a scientist, though, that doesn't explain why you should. After all, as a citizen of a modern society, you don't have to work out from first principles how to make an iPad or to program video games—somebody else has already done that. You could just pick up a tablet, download an app, and kick back playing Angry Birds while resting on the laurels of past generations of scientists. You don't need to be a scientist to make it in the world today. So why should you bother thinking scientifically?

On a practical level, making use of your inner scientist can make you more effective, both at generating eureka moments and at capitalizing on them once they occur. But the best reason to think like a scientist is that the scientific approach to the world is fundamentally optimistic and empowering.

It may seem odd to call science optimistic, given the regular stories about how scientists have discovered that we're all going to die in some gruesome manner—killed by drug-resistant bacteria, crushed by flaming rocks from outer space, roasted by global warming. I'm talking about a more abstract form of optimism, though. The scientific way of looking at the world is founded on the idea that the world is comprehensible; scientific thinkers know that questions have answers and that we can find those answers. In a very deep way, that's as optimistic as you can get.

Science is also empowering, because it gives us the tools we need to find answers to all sorts of questions. There are few things more frustrating than trying to deal with some problem and running into people who regard "I don't know" as a final answer. Scientific thinking turns "I don't know" into "I don't know . . . yet." The process of science lets you find answers to just about any question you might ask.

Your inner scientist is helpful not only for abstract matters of science, but also for almost any problem. If you are comfortable with the process of science, you can confront all sorts of situations with confidence. Even relatively mundane problems—figuring out how to use a new piece of computer software or deciding what process to follow for some task at work—can be addressed scientifically. You never need to stop doing something because you don't know some fact or don't know how something works. You can figure out whatever you need to know.

If you know how to use your inner scientist, you never need to settle for ignorance. And that's an incredibly powerful tool you can use to make your life better. It doesn't get more optimistic and empowering than that.

In this book, we'll dig into the process of science and look at how it applies to everyday activities. We'll not only show that you can act like a scientist, but also show how you already do, often without realizing it. All of us, from toddlers on, have a scientist inside us, and if you can recognize this capability and make use of it, your inner scientist will change your world for the better.

STEP ONE

LOOKING

The most exciting phrase to hear in science, the one that heralds new discoveries, is not "Eureka!" but "That's funny."
—Attributed to Isaac Asimov

The process of science, broadly defined, begins by looking at the world to identify something interesting that needs an explanation. This crucial first step takes many forms.

In observational sciences, the first step literally involves looking at the universe around us. Astronomers use telescopes to capture the light from distant stars. Geologists look at the layers of rocks exposed by erosion or in samples the scientists obtained by drilling into the ground. Biologists look for new organisms in nature or observe the behavior of species already known.

In experimental sciences, the process of looking involves deliberately perturbing in some way the subject being studied. A physicist bombards atoms and molecules with laser light and watches how they respond. A chemist combines various substances and sees what is produced when they react. A cognitive scientist places people or animals in unusual situations and looks at how their behavior changes.

Regardless of the field of science, the looking phase is a search for patterns. Observational scientists look for many examples of similar phenomena—astronomical objects emitting the same kinds of light, places with similar layers of rocks, animals that share particular traits. Experimental scientists look for repeatable responses to perturbations—atoms that always absorb a particular wavelength of light, chemical reactions that always produce the same products, persistent biases in people's responses to certain situations. The events studied can be exceedingly rare—some of the exotic particles studied by physics turn up only a handful of times in billions of particle accelerator collisions—but repeatability is essential. Singular events are not well suited to scientific explanation.

In this part of the book, we'll look at some examples of everyday activities that rely on the amassing of data and the search for patterns for success, and we'll give some historical examples of scientific discoveries that grew out of similar processes.

Chapter I

COLLECTING THE
ORIGIN OF SPECIES

All science is either physics or stamp collecting.
—Ernest Rutherford

One of the most straightforward hobbies is the collecting of things. If you think of an object that comes in more than one variety, you can almost certainly find somebody who collects it, and probably at least one Internet forum discussing the ins and outs of collecting whatever it is.

The urge to collect starts very young. Our daughter once spent many weeks assembling a rock collection. Of course, as she was four years old at the time, she wasn't exactly discriminating—most of her rocks were bits of gravel from the path to our backyard—but she took it very seriously. As any parent knows, toy makers have been quick to exploit this urge, churning out all manner of items—trading cards, stickers, stamps, plastic figures, cups, and plates—branded with images of popular characters, encouraging kids to collect every variation at several dollars a pop.

While many collectible items are targeted at children, millions of adults maintain collecting hobbies. Some of the items collected and traded are holdovers from childhood, such as baseball cards and collectible toys, which are snapped up as eagerly by adult collectors as by children. Others straddle the line between child and adult, such as comic books, sports memorabilia, and autographs. Still others are almost purely adult pursuits—I had a great-aunt who spent decades building a collection of china bells—and others collect china, crystal, porcelain figurines, and all manner of other fragile, decorative items. Some items are collected for intrinsic value, like art and antiquities—many excellent museums have their origin in this hobby—but for the most part, collectors accumulate things for the sheer pleasure of it.

Probably the most iconic collecting hobby is stamp collecting, which has been around long enough to get mentioned by the physicist Ernest Rutherford circa 1900 in his famous dig at other sciences (quoted at the start of this chapter).* The hobby is respectable enough to have a fancy Latin name.† It's a widespread and well-established hobby for a wide range of ages, with the American Philatelic Society listing more than five hundred local stamp collecting clubs in the United States and more than 650 registered dealers. Particularly rare or

---

* Amusingly, when Rutherford received a Nobel Prize for his discovery of the transmutation of elements through radioactive decay, the award was for chemistry rather than physics. He was a good sport about it, though, joking that "I have dealt with many different transformations with various periods of time, but the quickest that I have met was my own transformation in one moment from a physicist to a chemist." We'll talk more about Rutherford in Chapter 8.

† *Philately* is properly defined as the study of stamps, but many stamp collectors consider themselves philatelists. The door of the lab where I worked in graduate school had a poster bearing this Rutherford quote; in response, another scientist in the building changed the official door label to read "NIST Philatelics Laboratory."

important stamps have sold at auction for millions of dollars. Famous stamp collectors include heads of state (US president Franklin Roosevelt and King George V of Britain), authors (Ayn Rand), and entertainers (Freddie Mercury and John Lennon both had childhood stamp collections, now held by postal museums in the United Kingdom and the United States, respectively).

Like a lot of kids, I had a stamp collection for a while. I never collected anything particularly notable, but going through old letters and boxes of stamps from relatives who had had collections was enjoyable in a quiet way. And putting the individual stamps together to make a larger picture was fascinating. I remember an intimidatingly large three-ring binder with spots for every US stamp that had been issued to that point, and the satisfaction of completing a page. My hobby also gave a sense of history outside the collection—for example, seeing all the stamps of the 1893 Columbian Issue commemorating the 400th anniversary of Christopher Columbus's famous voyages showed me there was a good deal more to the story than I had heard in grade school.

Beyond the immediate pleasures of building a collection, though, the impulse to collect can be a starting point for science. The most obvious product of collecting hobbies is an array of physical objects, but collecting is also a mental state. Serious collectors develop habits of mind particular to their hobbies—a sort of constant low-level awareness of possible sources of stamps, an ability to spot new specimens, and close observation and knowledge of the fine gradations that separate valuable stamps from worthless bits of colored paper. These habits of mind also serve well in science; the simple act of collecting a diverse array of interesting objects or observations also serves as the starting point for most sciences.

Rutherford's famous gibe contains a small element of truth, because the physics of his day was more fully developed than other sciences, in terms of successful unifying theories like Newton's laws of motion and Maxwell's equations for electromagnetism. But that very development started with the "stamp collecting" of lots of individual bits of data. Newton would not have been able to formulate his laws without decades of carefully recorded astronomical observations and experimental tests by previous generations of scientists. Maxwell's equations bring together the results of dozens of seemingly unconnected experiments on the behavior of charged particles and magnets. And the amassing of examples continues to be critically important to this day—the Standard Model of particle physics is arguably the most comprehensive and successful scientific theory in human history, but modern particle physics is the ultimate big-data science, with the Large Hadron Collider (LHC) producing hundreds of petabytes worth of experimental data in a year.*

The other sciences at the turn of the twentieth century were nowhere near as fully developed as they are now or as physics was then. In geology, the idea of continental drift was decades off and scientists still had some trouble determining the age of the Earth—the best available estimates from the temperature of the Earth and Sun suggested an age of at most a few hundred million years, a small fraction of the apparent age from rocks. In chemistry, the rules determining bonding of atoms into molecules were known, but the underlying principles were not understood until the development of quantum mechanics; there was even some debate as to whether atoms were real physical entities or merely a mathematical

---

* One petabyte is 1,000,000,000,000,000 bytes.

convenience. Biology was probably the furthest along, but even there, the rules of heredity were still being worked out and the discovery of DNA as the mechanism of heredity, one of the crucial foundations of modern biochemistry, was nearly a half-century away.

All of these sciences have made remarkable progress in the past century, matching or even surpassing the development of physics in Rutherford's day. The development of our modern understanding of all of these sciences began with the collection of a huge number of "stamps," allowing scientists to determine patterns that are only clear through the accumulation of a vast array of information. So while there is some truth to Rutherford's snide taxonomy, in another sense, it misunderstands the process of science. Stamp collecting is an essential step on the way to deeper scientific understanding. This is best illustrated by what may well be the most important and controversial scientific book ever written, Charles Darwin's *On the Origin of Species*.

FROM *BEAGLE* TO BOOK: DARWIN'S BACKGROUND

In December 1831, HMS *Beagle*, under the command of Captain Robert FitzRoy, set sail from England to perform a hydrographical survey of South America. The intended mission was fairly unremarkable—it was a follow-up to an earlier survey (now largely forgotten) conducted from 1825 to 1830. Accompanying the ship was a young man from a prosperous family, Charles Darwin, who was himself fairly unremarkable. Darwin had recently completed his education at Cambridge and signed on to the *Beagle* as a sort of unofficial naturalist and dinner companion for FitzRoy. Darwin's father initially disapproved of this "wild scheme," but relented after the intervention of his brother-in-law, Josiah Wedgwood. Charles took ship for five years to make a study of the geology, plants, animals, and people of South America as the *Beagle* picked its way down the eastern coast of South America, around Tierra del Fuego, and up the western coast of South America before striking off across the Pacific to return to England.

It's hard to imagine that anyone involved could have known how momentous this voyage would turn out to be. And indeed, for twenty years the trip didn't appear to have been particularly earth-shaking. Young Darwin acquired a huge number of specimens of interest to science, including quite a few new species and some impressive fossils, but his work initially consisted of just picking up whatever interesting-looking items the expedition happened to encounter, without any grand theoretical goal. One of his new bird species, a small variety of rhea, was discovered when Darwin realized the ship's company was eating it for dinner.* And when collecting samples of the many finch species of the Galapagos Islands, Darwin initially didn't bother to record which birds came from which island.† He wouldn't have recognized the reference to stamp collecting, as the first postage stamp wasn't issued until 1840, but Darwin on the *Beagle* started out operating very much in the "stamp collecting" mode.

The primary immediate impact of the voyage was personal, confirming his passion for the study of nature. Darwin wrote a popular book about his travels as part of a series about the voyage (the *Journal of Researches*, now known as *The Voyage of*

---

* He preserved bits of the carcass to send back to England and later saw and described the new species in the wild.

† Darwin's finches have since become an iconic representation of evolution, thanks to the adaptation of their beaks to the different food available on each island, but this was only through the work of later naturalists, chiefly David Lack in the 1940s.

the Beagle), and between the success of the book and Wedgwood family money, he was able to abandon the original plan for him to join the clergy.[‡] In 1842, he settled down in a country house to study nature at his leisure.[§] And the voyage did plant seeds that would lead to his later work—while he didn't carefully record the origin of the Galapagos finches, he did note striking differences between the mockingbirds on different islands, which started him questioning whether divisions between species were as sharp and immutable as scientists then thought.

Although he never needed to work at a trade, Darwin was hardly idle, studying a wide range of organisms in great detail and so obsessively that his children assumed that all fathers spent their days peering through microscopes. One of his sons, visiting a friend's family, famously asked, "Where does your father do *his* barnacles?" His books about barnacles won a prize from the Royal Society, and he was an active member of the leading scientific societies of the day. He also carefully tracked the behavior of plants in his gardens and spent many years raising pigeons.

The fruit of all this labor came in 1859, when Darwin published his most important book, *On the Origin of Species by means of Natural Selection, or the Preservation of Favoured Races in the Struggle for Life.* This is the book that launched the modern concept of biological evolution, sparked enormous public controversy, and made Darwin a figure to be reviled or revered by various combatants in the culture wars that continue to this day.

[‡] Josiah Wedgwood, the uncle who intervened to get him permission to voyage with the *Beagle*, later became Charles's father-in-law and was heir to a fortune made in the pottery business.

[§] This was fairly common at the time; most of the great names of British science up through the 1800s were wealthy gentlemen who pursued science as a sort of hobby, the way others collected art or antiquities or even stamps.

Darwin was not, however, the first person to think in evolutionary terms—not by a long shot. His own grandfather Erasmus Darwin, a physician and poet, had been promoting evolutionary ideas in the late 1700s, and the French scientist Jean-Baptiste Lamarck had worked out his own detailed theory of gradual change in animal species beginning around the same time. Darwin was enormously influenced by another Charles, the geologist Charles Lyell, whose *Principles of Geology* he read on the *Beagle*. Lyell argued that rather than being created at some specific point within the last ten thousand years or so, the Earth had existed for an unimaginably long time, with its geology slowly but steadily changing due to slow processes of uplift and erosion.*

Why, then, does Darwin have iconic status, while earlier evolutionary thinkers are only remembered by historians of science? Darwin's *Origin* supplanted earlier theories for two reasons. First, he provided a clear mechanism by which evolutionary changes occur: the slow accumulation of small variations that make specific individual organisms more likely to survive and reproduce. Those beneficial changes are passed on to future generations, where further variations occur, with the beneficial changes passed on, and so on.

This mechanism of natural selection places Darwin's theory squarely within the realm of science.† Erasmus Darwin and Lamarck had invoked vague metaphysical "principles"

---

* Lyell did not apply the same reasoning to biological systems and in fact devoted some space to attacking the evolutionary theories of Lamarck.

† By removing the need for any intelligence guiding the process, natural selection also presents a dramatic challenge to religious ideas of divinely guided creation, creating instant controversy from the day of its publication right down to the present. Although the furor was personally uncomfortable for Darwin, it certainly helped cement his status as an icon of science.

driving continual improvements in species, as if there were some goal to be attained, which is an uncomfortable basis for a scientific theory.‡ Lamarck provided the first coherent theoretical framework for talking about evolution, and Darwin himself cited it as an influence on his thinking, but without a better idea of the mechanism of change, Lamarck's theory gained little traction as science.

The second, equally important, factor that secured Darwin's triumph came from his years of careful collecting, from the *Beagle* to his barnacles to the breeding of pigeons. The *Origin* succeeded as brilliantly as it did in large part because Darwin supported his argument with concrete evidence, piles and piles of it. All those years of collecting and cataloging plants and animals like stamps in an album paid off.

INSIDE THE *ORIGIN*

Despite the lengthy title, *On the Origin of Species by means of Natural Selection, or the Preservation of Favoured Races in the Struggle for Life* is fairly compact, particularly by the standards of Victorian literature—around five hundred pages, in a single volume.§ Darwin himself thought of it as an abstract for a much longer work, which he never completed, and refers to it as such within the text.¶

---

‡ Another pre-Darwin evolutionary work, the sensationalist *Vestiges of the Natural History of Creation*, published anonymously in 1844, compounded these metaphysical "principles" with wild speculations that were implausible even in Victorian times. It became a best seller, but the blistering criticism it drew from scientists may have made Darwin more hesitant to publish his own ideas.

§ The *Origin* went through six editions in Darwin's lifetime, involving many revisions and additions to the text. In this book, I refer to the first edition, specifically, the electronic version of the Barnes and Noble Classics edition.

¶ The *Origin* was rushed into print after Darwin received a letter from the

Those pages are packed with information, though. Darwin's relatively breezy and conversational style flows smoothly enough to be an easy read even today (though the lengthy and complicated sentences characteristic of Victorian prose take a little getting used to), but he presents a mountain of evidence, from a dizzying array of sources. He illustrates his points with references to all manner of plants and animals, cites the research of eminent scholars and conversations with farmers and animal breeders, and describes the results of his own experiments. Hardly a paragraph goes by without at least one reference to specific observations of the natural world.

This vast array of information is necessary because convincing people of evolution through natural selection is an inherently difficult task. By its very nature, the process of evolution is extremely slow.[*] Evolutionary change requires many generations—probably hundreds—to become visible. The process is not easily observed directly—it would require either organisms with exceptionally short lifetimes (so that human observers could watch enough generations go by) or the ability to look back over thousands if not millions of years.

Since the changes happen so slowly, an argument for evolution, particularly in 1859, must be more circumstantial. We are able to directly observe plants and animals only in a

---

naturalist Alfred Russel Wallace outlining a very similar evolutionary theory. Although Darwin had been working on a manuscript as early as the 1840s, he discarded it and started over, producing his most famous book in just thirteen months.

[*] The specific result is also difficult to predict, because of the inherent randomness of the variations that drive the process. In general, future generations will be better adapted to their new environment, but there are many ways to accomplish that adaptation, and the results are sometimes surprising.

narrow slice of time (the past few hundred years, at most), with fragmentary additional evidence from fossils created in the more distant past. We can see traces of evolutionary history in the structure and behavior of modern plants and animals, however, and while the evidence from a single species or group of organisms may not be terribly convincing by itself, the accumulation of evidence from many different organisms of different types produces a compelling argument.

Darwin's argument in the *Origin* follows a straightforward progression. First, he establishes that small variations in the characteristics of different organisms occur randomly in nature and that these characteristics are inherited by the next generation. Then he argues that all living things are constantly engaged in a "struggle for existence," competing with other organisms and against environmental factors for the resources needed to survive and reproduce. Given those two factors, the idea of natural selection follows directly: The random variation of individual characteristics will occasionally produce individuals with some trait that makes them more likely to survive and reproduce—longer legs to better outrun predators, say, or a longer snout to better extract food from hard-to-reach places. They will pass this trait on to their descendants, some of which will have still further variations—even longer legs or snouts— that make them more likely to survive and reproduce, in turn passing their variations on to future generations, and so on. The changes from one generation to the next will be too small to notice, but over time, they will accumulate, eventually producing organisms with characteristics dramatically different from those of their ancestors. In a place where predators are particularly common, a speedy and long-legged version will be found, while somewhere else, where finding food is more important than avoiding predators, a long-snouted version might dominate. This is natural selection, an analogy to the deliberate selection employed by breeders of domestic plants and animals. Plant and animal breeders allow only individuals with desired characteristics to reproduce, eventually producing very distinct breeds. Unlike domestic breeding, however, natural selection occurs automatically, with no particular intent driving it.

Each step of this argument is backed up with a huge amount of evidence from different species of animals and plants. Darwin addresses variation in two dense opening chapters, the first establishing variability and inheritance among domesticated animals, and the second arguing for similar variability in wild animals. The first chapter alone cites observations regarding ducks, cows, goats, cats, sheep, pigs, dogs, cattle, donkeys, guinea fowl, reindeer, camels, rabbits, ferrets, horses, geese, and peacocks, along with at least ten named breeds of pigeons, to illustrate the effects of animal breeding. The chapter goes on to mention gooseberry, cabbages, roses, pelargonium, dahlia, and pears (along with references to fruit and flowers of unspecified varieties) to show the influence of horticulturists. He also cites work from at least twenty-four sources, ranging from other scientists and noted animal breeders, to the Roman writer Pliny and the book of Genesis. All of this while regularly apologizing for being unable to go into sufficient detail in the space of a single book.

Having established that individual organisms vary in their characteristics and that these variations are passed on to their offspring, Darwin moves on to establish the struggle for existence. The argument here is, as he puts it, "the doctrine of Malthus applied with manifold force to the whole animal and vegetable kingdoms." The reference is to the Rev. Thomas Malthus, whose 1798 *Essay on the Principle of Population* argued that human population will necessarily grow until checked by famine or disease.* Darwin extends this idea to all of nature, arguing that plants and animals inevitably reproduce at rates that would quickly exhaust the available resources to support them, if not for other forces acting to reduce their populations.

Again, this step of the argument is copiously illustrated with examples from his own work and that of others. He includes a calculation estimating that a single pair of slow-breeding elephants reproducing unchecked would produce fifteen million descendants in five hundred years; as we're not all in danger of being trampled, there must be something checking the elephant population. He describes an experiment showing that fast-growing weeds will overwhelm other plants when a cleared patch of ground is allowed to fill back in. And he reports an accidental experiment by one of his country neighbors, where enclosing a tract of land to keep animals out led to a dramatic increase in the abundance of Scotch pines, whose seedlings were otherwise eaten by grazing livestock. These observations clearly demonstrate that plants and animals compete with each other for available resources, that small changes alter the balance of that competition, and that altering the balance leads to readily observable changes in the populations of the various species present.

Darwin's fourth chapter puts these ideas together to introduce the core idea of evolution through natural selection: Since individual organisms are prone to vary in their characteristics and must struggle to avoid starvation or predation, any variation that increases the likelihood of success in that struggle will be more likely to be passed on to the next generation.

---

* This cheery observation is a big reason why Malthus's field, economics, is nicknamed "the dismal science."

This is necessarily the most abstract chapter of the book, relying in large part on hypothetical arguments about the possible course of evolution for various organisms. But even Darwin's hypotheticals come with citations and evidence. After considering the operation of natural selection on a population of wolves hunting deer, he immediately cites the observation of two distinct types of wolves in the Catskill Mountains of New York, as suggested by his hypothetical model. After considering the operation of natural selection on hypothetical flowering plants and insects, Darwin reports the results of his own observations of plants in his garden, showing the effects of flower structure on their fertilization by insects, which again fit with his hypothetical model.

Like a good student of rhetoric, he devotes much of the book—seven of the fourteen chapters—to anticipating and addressing objections to the theory, backed up with still more observations and experiments consistent with the theory. Many of the apparent "difficulties" Darwin notes—the apparent improbability of evolving an organ as complex as an eye, for example, or the apparent lack of "transitional forms"—are still cited in anti-evolution propaganda, but none pose any real obstacle to the theory. Darwin's counterarguments are backed with the best evidence available to him and in many cases set the pattern for modern responses to the same questions (these days backed with still more evidence from the intervening 150 years).

While none of the individual bits of data are decisive, his encyclopedic array of examples, spanning much of the globe and extending to all corners of the animal and plant kingdoms, produces a convincing and formidable argument. It doesn't hurt that he is also a charming and persuasive writer, but the *Origin* would not be the book it is without the twenty-plus years that Darwin spent collecting "stamps" to serve as examples in support of his argument.

EVOLUTION SINCE THE *ORIGIN*

In the 150 years since Darwin published the *Origin*, the already strong case supporting his theory has become overwhelming, and evolution by natural selection is the cornerstone of modern biology. Although Darwin's theory was controversial when it was published, the geneticist Theodosius Dobzhansky published in 1973 a famous essay titled "Nothing in Biology Makes Sense Except in the Light of Evolution," a phrase widely adopted by evolutionary biologists. Dobzhansky argued that without the unifying concept of evolution, biology would be nothing more than "a pile of sundry facts, some of them interesting or curious but making no meaningful picture as a whole"—just a stamp collection, in Rutherford's terms. With evolution, everything fits together, across a diverse collection of subfields; paleontology, genetics, cell biology, and ecology all provide clear evidence of evolution.

Essentially all of the "difficulties" Darwin identified with his theory have been addressed. In some cases, finding answers was simply a matter of further time and research. Darwin noted that the fossil record of the history of life is necessarily somewhat patchy, as the conditions needed to preserve fossils for millions of years are rare. A century and a half of further fossil collecting, though, has provided innumerable examples filling in gaps in the record of Darwin's day, clearly tracing the slow change of species over millions of years of evolution. New evidence continues to come to light, with recent discoveries like the limbed fish *Tiktaalik* shedding new light on the transition from purely ocean-dwelling fish to land-dwelling tetrapods.

Some other scientific advances have made it possible to directly see evolution in action. The germ theory of disease had not yet caught on when Darwin was writing, and antibiotic drugs were still decades off, but the modern medical problem of antibiotic resistance clearly demonstrates the effects of natural selection in the wild. Disease-causing organisms are more likely to survive and reproduce if they have some small variation that makes them less susceptible to the antibiotics doctors and hospitals use to treat infections. If that variation is inherited by the organisms' descendants, these resistant off-spring will eventually dominate the population of infectious organisms. Drugs that used to be effective at killing particular bacteria are observed to become less and less effective over time, a real and deadly phenomenon that can only be understood in terms of evolution by natural selection.

In other cases, new scientific advances allowed biologists to put Darwin's theory on more solid conceptual ground. In the *Origin*, Darwin is forced to admit that the exact rules of inheritance are a mystery, and he struggles to explain exactly why individuals of a species vary in their characteristics and how their traits are passed on. Beginning with the plant studies of Gregor Mendel, an obscure contemporary of Darwin's whose work was rediscovered in the 1900s, biologists developed a concrete and quantitative understanding of inheritance. With the development of biochemistry and the discovery of the DNA code, our knowledge of heredity exploded. Modern biologists have a vastly better understanding of how genes combine to produce individual traits and how small errors in copying genes (or damage by environmental factors) can introduce the new variations needed to drive evolution by natural selection. Hardly a week goes by now without some splashy press release touting the discovery of a gene associated with some trait or another, and some biologists regard genes, not organisms, as the central elements when talking about evolution.

The evidence for evolution is so broad and so overwhelming that I'm forced to take a cue from Darwin himself and apologize for lacking the space to do justice to so complicated a subject. There are countless books and websites devoted to the subject for those interested in reading more, particularly the excellent Understanding Evolution site run by the University of California Museum of Paleontology (http://evolution .berkeley.edu/evolibrary/home.php), which includes articles detailing the very latest developments in evolutionary biology.

Some modern discussions of evolution, particularly in the area of genetics, would be all but incomprehensible to Darwin, while others would seem comfortingly familiar. One thing, though, has remained unchanged since Darwin's time: The case for evolution through natural selection is not sealed by any single spectacular piece of evidence, but by an immense collection of small clues pointing to the same answer. Science cannot wind evolution backward like some home movie, showing a seamless transition from humans to protohumans to chimpanzee-like creatures, to tiny early mammals living in the shadow of dinosaurs. That level of detail will never be possible. But the vast preponderance of evidence, collected together, one piece at a time like stamps in an album, points unmistakably to the evolutionary origins of humans and, indeed, of all living things today.*

BEYOND THE *ORIGIN*

While Darwin's work is probably the highest-profile example of clinching a discovery through the amassing of vast piles

---

* Sadly, there are still people who refuse, on religious grounds, to accept this evidence, and they oppose the teaching of evolution in schools. Their arguments are entirely without merit and wouldn't deserve even a footnote in a book about science were it not for their well-funded and politically connected allies, particularly in the United States. As a result, constant vigilance is required from scientists and organizations like the National Center for Science Education (http://ncse.com) to protect the integrity of science education.

of evidence, all sciences start with the collection of "stamps," small bits of data that may seem no more than faintly interesting curiosities at first glance. This is particularly true of the observational sciences, where researchers attempt to piece together long-ago events that are not easily repeated. The idea that the continents shift position over time began with Alfred Wegener's observation that the coastlines of Africa and South America seem almost like complementary puzzle pieces. The idea of continental drift didn't gain acceptance until after multiple lines of other evidence were found to support it: close similarities between rock strata and fossils on opposite sides of the Atlantic, evidence of sea-floor spreading at the Mid-Atlantic Ridge, and matching "stripes" of magnetization in rocks on either side of the ridge, tracing out the history of magnetic pole flips. Thanks to the convergence of all those bits of evidence, the idea that the continents shift position over millions of years is central to the modern theory of plate tectonics, as central to geology as evolution is to biology.

In astronomy, the Big Bang theory of cosmology has similarly attained a nearly unassailable position at the heart of our understanding of the universe, through the accumulation of many lines of evidence. Vesto Slipher and Edwin Hubble noted that distant galaxies all appear to be receding from us, as you would expect in an expanding universe. And Arno Penzias and Robert Wilson discovered the "cosmic microwave background radiation," a remnant of light left over from when the universe was only 300,000 years old. The existence of this background radiation had been predicted by Ralph Alpher and Robert Herman as a necessary consequence of the Big Bang. And the mysterious dark matter and dark energy that account for some 96 percent of the energy content of the universe are accepted by the vast majority of scientists, despite this stuff's never having been directly detected. But thanks to the convergence of many observations that suggest the presence of vast amounts of matter we can't see (we'll discuss some of these observations in Chapter 6), scientists accept the presence of dark matter and dark energy. Like Darwin's finches and fossils, none of these observations are conclusive by themselves, but all of them together make a case that only a tiny minority of die-hards continue to deny.

Even in experimental sciences, though, the collection of evidence from multiple sources plays an essential role. As we'll see in Chapter 8, our modern understanding of atoms was pieced together from hints seen in many experiments over a period of decades—with some of those "stamps" collected by Rutherford himself.

Rutherford's quip about stamp collecting is usually brought out as either a teasing dig at biologists or an example of the overbearing arrogance of physicists, depending on whether the person citing it is a physicist or a biologist. But given the essential role of data collection to evolutionary biology and other sciences, perhaps Rutherford's comment should be viewed more as a compliment to stamp collectors. Nearly all progress in science can ultimately be traced to the human impulse to collect and arrange enormous amounts of stuff.

STAMP COLLECTING AND PROBLEM SOLVING

We usually think of collection hobbies as an end unto themselves. Most stamp collectors are not building their collections in service of a larger theory, but are doing so because they enjoy gathering stamps together and filling gaps in their collection. As these stories from the history of science illustrate, though, the collecting impulse can be a starting point for scientific progress. The habits of mind that make a good collector—always being on the lookout for new specimens, noting small and interesting features, a drive toward completing the set of all possible variations—also work to the benefit of scientists.

The collecting impulse can also play an important role in solving problems we wouldn't usually call scientific. When our daughter was a baby, she was prone to prolonged episodes of inconsolable screaming and crying, associated with pain in her stomach. This was officially diagnosed as colic, a medical term meaning "terrible stomach pain for no obvious reason." Kate and I were assured that this was a common problem that would go away on its own within a few months. When it didn't go away, we eventually figured it out through stamp collecting.

We had already been keeping careful track of what our daughter ate and when, because that's our general inclination (and I had been making colorful charts and graphs out of this information, because that's what I do). Putting together many weeks of observations of her diet and when her stomach acted up, we began to see a pattern, and adding in records of what Kate was eating and drinking made everything come clear. The root cause of the problem was that our daughter was sensitive to both soy and dairy, including the soy and dairy that Kate ate while she was breastfeeding.

None of the individual records were all that useful, and most of the simple experiments we tried were failures. Switching to nondairy formula for supplemental feedings didn't help, because the nondairy formula contained soy, which was also a problem. Nondairy, nonsoy formula didn't work as well as expected, because Kate was still eating dairy products. It was only when we put all the separate observations together that the true pattern became clear. Once we phased out all the soy and dairy from everyone's diet, though, the colic problem finally resolved itself, and we were all able to get some sleep.*

The story of our colic adventures illustrates an important point about the process of science. For a scientific approach to be effective, you need to have data. Often, the individual measurements that make up that data will seem insignificant, even trivial, like individual postage stamps. Put together enough stamps, though, and you can have something very valuable.

The lesson to take from Darwin and Rutherford and collecting generally, then, is the importance of measuring everything. The first step to bringing your inner scientist to bear on a problem is to collect as much information as you can about the problem—if you want to lose weight, you need to track what you eat; if you want to make better use of your time at work, you need to track what you do through the day. The individual records may not seem meaningful in themselves, but taken all together, they may reveal useful patterns and suggest solutions.

---

* The elimination-diet process is not as simple as it might sound, because there's soy in all sorts of places you wouldn't expect. Happily, however, the sensitivity to soy and dairy eventually went away, and cheese and yogurt now account for roughly 80 percent of the calories in our daughter's diet.

Chapter 2

SCIENTIFIC CUISINE
REIGNS SUPREME

From 1999 to 2001, Kate and I lived in New Haven, Connecticut, where she was a law student and I was a physics postdoc. These are both pretty intense occupations, so on Friday nights, we tended to just hang around my apartment and watch television. One of the staples of our TV-watching was a then-little-known cooking competition from Japan, *Iron Chef*, airing for the first time on the Food Network. The premise of the show was simple: Each week, an accomplished chef would challenge one of the show's stable of Iron Chefs to a battle, with each chef getting one hour to prepare dishes from a secret theme ingredient unveiled at the start of the hour.

Originally airing from 1993 to 1999 in Japan as *Ryori no Tetsujin* (which translates literally as something like "Iron Men of Cooking"), the show became a cult hit on US cable. It had an abundance of campy charm: the flamboyant Chairman Kaga Takeshi, the play-by play announcer Fukui Kenji, the hardworking floor reporter Ohta Shinichiro, and the avuncular color commentator Hattori Yukio; the pomp and circumstance of the chefs entering Kitchen Stadium; and even the array of cranky judges who always seemed to give a narrow victory to the Iron

Chef. Food Network has been airing *Iron Chef America* since 2005, but while the format is similar, the American version is a pale imitation of the original, mostly because it's too knowingly ironic. As over the top as the Japanese version was in some ways, everyone involved appeared to take it very, very seriously.

Of course, the real highlight of the show was the food, which was always entertainingly exotic. Sometimes this unique flavor was merely a function of the show's Japanese origins—ingredients like shark's fin, *konnyaku* (a sort of gelatinous starch product), or *natto* (fermented soybeans) are thoroughly baffling to Americans. More often it was due to ostentatious expense—it was a rare battle where neither of the chefs broke out the foie gras and truffles, and they always delighted in relating the total value of the more expensive theme ingredients. On one memorable show, a challenger used a half-dozen expensive lobsters just to add flavor to an asparagus dish, discarding the lobsters without serving them. But mostly it was a matter of unusual culinary combinations that no home cook would think to attempt—cooking food with a blowtorch, marinating seafood in champagne, or making fish-flavored ice cream.

If you watched a lot of the show, though—and we watched a lot of it in New Haven—some patterns began to emerge. Most of the Iron Chefs had go-to techniques: Iron Chef French Sakai Hiroyuki would almost always make some sort of intricate salad in a ring mold, and Iron Chef Chinese Chen Kenichi would invariably make a stew involving chili peppers. Patterns among the challengers were harder to spot, but again, there were certain techniques you could count on seeing almost every week.

That's the relevance of *Iron Chef* for this book.* One of the

---

* Which is not to say that there isn't explicit science involved in cooking— there's plenty of science in the kitchen. Another Food Network staple,

defining characteristics of good cooks is that they know techniques rather than just recipes. Almost anybody can follow a recipe, given all the ingredients and sufficiently clear instructions, but a good cook can take whatever he or she has on hand and find a way to make something delicious. The key to cooking without a recipe is knowing a few basic methods and successful combinations of flavors that can be applied to a wide range of ingredients. If you understand how to roast potatoes, you can probably manage to cook other root vegetables or squash. If you have a marinade that works well when grilling steak, you can almost certainly use it with other red meat like lamb or venison as well.

The importance of techniques rather than specific recipes is something that a lot of novice cooks fail to understand. It's also something some of my favorite cookbooks—Alton Brown's *I'm Just Here for the Food*, James Peterson's *Essentials of Cooking*, and Mark Bittman's *How to Cook Everything*—address explicitly, organizing themselves around cooking techniques and explaining how to apply a few methods in many different dishes.

This mode of operation also carries over into science, particularly experimental science. Good experimentalists generally know a few tricks that they then apply to a wide range of problems, and progress in science has often come from applying a method from one field to a problem in a different field. Something similar happens in theoretical science, though not quite as frequently—Albert Einstein is best known for the theory of relativity, but his training was in statistical physics, and he made important contributions to understanding the nature of light by applying statistical methods to the new theory of quantum physics.

Alton Brown's *Good Eats*, had a twelve-year run explaining the science behind a lot of ingredients and techniques.

By way of illustration, we'll take a look at the career of one particular scientist, the physicist Luis Alvarez.* Like a lot of his contemporaries in physics, Alvarez led a fascinating life, with his career spanning the development of quantum mechanics, the early glory years of nuclear and particle physics, and the massive scientific projects of World War II. His autobiography is a terrific read.†

His work in experimental particle physics earned him the 1968 Nobel Prize in Physics, but he's arguably most famous for the contributions he made to other fields. In particular he led an experiment to "x-ray" one of the pyramids at Giza and helped discover an asteroid impact that happened sixty-five million years ago and that may have wiped out the dinosaurs. While these projects seem very far removed from his home field of particle physics, they draw on expertise he developed in physics, and the projects succeeded because he brought tools and tricks he learned from his training as a physicist into a new context.

COSMIC RAYS AND THE PYRAMIDS OF GIZA

The first of Alvarez's big accomplishments outside physics has its roots in 1932, when he was still an undergraduate at the University of Chicago. For a senior research project, he built two Geiger counters—the iconic radiation detectors of the cold war,

---

* I have a personal reason for picking Alvarez as my example, beyond the facts of his biography. In 1981, when I was in fourth grade, I wrote him a letter asking some questions about his theory that an asteroid impact killed the dinosaurs. He wrote an extremely gracious reply, treating my questions very seriously, which made a huge impression on me.

† Luis W. Alvarez, *Alvarez: Adventures of a Physicist* (New York: Basic Books, 1987).

that make clicking sounds when a radioactive particle passes through a sealed tube. The instrument had only just been invented in Europe by Hans Geiger, so Alvarez's counters were the first built at Chicago. By his own account, they weren't very good, often clicking even when no radiation was present.

Arthur Holly Compton, one of Alvarez's mentors at Chicago, suggested a way to make these noisy counters useful. While individually they produced many spurious clicks, the two could be combined to form a "cosmic-ray telescope" using a technique known as coincidence counting. The basic scheme is shown in Figure 2.1. Alvarez took the output of the two counters and fed it into an electronic circuit that would produce a click only if both counters recorded a click at very nearly the same instant. The spurious clicks of the individual detectors occurred randomly, and the chance of both tubes happening to click at once was very slim. A signal that caused both detectors to click at the same time was almost certainly real, due to a radioactive particle passing through both detectors in rapid succession.

Figure 2.1. Diagram of a cosmic ray telescope based on coincidence counting.

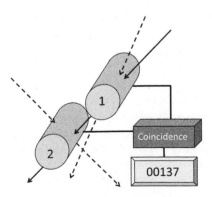

This coincidence-counting arrangement would only detect radiation coming from a small range of directions. A particle coming along just the right line, represented by the solid arrow in the figure, will hit both detectors and produce a click that can be fed into an electronic counter. Particles coming from other directions, represented by the dashed arrows in the figure, may hit one of the two detectors, but will not hit both and thus will not be counted.

The most likely source of a particle passing through the two detectors is a cosmic ray, one of the huge number of charged particles from outer space constantly bombarding the Earth. Alvarez's linked Geiger counters function like a telescope for these cosmic rays, measuring the number coming from whatever direction it is pointed.

As Compton suggested, Alvarez built the coincidence circuit and then used it to make his first major discovery. In the 1930s, there was still a good deal of debate about what, exactly, the cosmic rays were, in particular whether they were high-energy light (gamma rays) or electrically charged subatomic particles. After years of argument, physicists realized that the Earth's magnetic field provided a way to distinguish between these cases: The neutral gamma rays would be unaffected by the field, but charged particles would feel a magnetic force pushing them to one side or the other, with positive charges curving in from west to east and negative charges curving from east to west.* This bending should produce a difference

---

* Low-energy cosmic ray particles are, in fact, bent enough to miss the Earth's surface altogether. The Earth's magnetic field thus serves as a shield protecting us from potentially deadly radiation. This is one of the factors that makes long-term space travel (such as a mission to send humans to Mars) such a dangerous and expensive business. A trip to another planet means leaving the Earth's protective magnetic field, either requiring heavy radiation shielding or running a serious health risk.

in the number of cosmic rays detected coming from the east and west: Negatively charged particles should come in from the east in greater numbers, while positively charged particles should come in from the west in greater numbers.

Alvarez's cosmic ray telescope provided a way of testing this, so at Compton's suggestion, he spent a month in Mexico City measuring the number of cosmic rays coming from various directions.[†] He mounted his telescope to the hinged lid of a box so that he could change the angle of the telescope, and he put the whole contraption in a wheelbarrow, which he pushed around to reverse direction from east to west every half hour. Both Alvarez and another American physicist, Tom Johnson from Swarthmore College, found a clear difference in their signals, with many more particles coming from the west than the east. This was the first clear proof that cosmic rays are predominantly positively charged; today, we know that they are primarily high-energy protons originating outside the solar system. Alvarez and Compton wrote up their results and published them in the *Physical Review*, the first of many significant papers in Alvarez's long career.

Alvarez eventually won the 1968 Nobel Prize for developing the hydrogen bubble chamber, which for decades was the dominant detector technology in particle physics. The idea of coincidence counting—looking for events where the detectors record two or more particles of particular types—is central to experimental particle physics, and Alvarez was involved in many discoveries that hinged on seeing these coincidences. Thirty-five years after he was introduced to the idea by Compton, this same technique also let him settle a long-standing question in archeology.

[†] The expected signal for this experiment is larger at high altitude and lower latitude, and Mexico City offered a convenient location with both.

The three pyramids at Giza in Egypt are among the grandest monuments ever built by humans. The largest of these, the Great Pyramid of Khufu, contains a complex network of tunnels and chambers, but the second-largest, the Pyramid of Khafre, contains only a single chamber near the center of the base. The simplicity of Khafre's pyramid has struck many people as improbable. Among the doubters was Luis Alvarez when he first visited Giza in 1962: "If [Khafre's] grandfather built two chambers and his father three, it seemed most likely to me that [Khafre] would have ordered four."[*]

Of course, any chambers that might exist within the pyramid were hidden from view by thousands of tons of rock. Finding a secret chamber inside the pyramid would require some way to see through all that rock. Like a Szechuan chef reaching for chili peppers, though, Alvarez realized he could use the familiar techniques of particle physics to answer this question. A scaled-up version of his undergraduate cosmic-ray telescope could use cosmic rays to "x-ray" the pyramid.

Some of the vast numbers of cosmic rays striking the upper atmosphere of the Earth collide with air molecules and produce particles known as muons, which are closely related to electrons, but much heavier. Muons are unstable but relatively long-lived and fast-moving, so many of them survive to reach the surface of the Earth and even some distance below it—high-energy muons can travel through hundreds of meters of solid rock.[†] Some of these muons reach all the way into the

---

[*] Alvarez, *Adventures of a Physicist*, 230.

[†] For this reason, astrophysics experiments looking for rare particles are often located deep underground, in abandoned or even working mines (see Chapter 3). Having several hundred meters of rock above the detector blocks most of the muons that might otherwise swamp the rarer particles of interest.

central chamber of Khafre's pyramid. The exact number of muons getting through depends on the thickness of rock they pass along the way, so measuring the number of muons coming from a given direction gives a measure of the thickness of the pyramid along that line. Measuring this thickness along many different directions and comparing it to what you would expect from a solid pyramid should reveal any hidden chambers. If more muons show up from a given direction than expected, that means they passed through less rock than expected, indicating a hollow space within the pyramid: a hidden chamber.

Together with collaborators in the United States and Egypt, Alvarez put together the apparatus shown in Figure 2.2 and installed it in the chamber at the base of Khafre's pyramid in 1967. The apparatus consisted of a multilayer sandwich of particle detectors: Starting at the top, there was a trigger detector that would make an electrical pulse when a muon passed through any part of it, then two spark chambers, then a second trigger detector, then a massive slab of iron, and a third trigger detector at the bottom.

Figure 2.2. The stack of detectors that Alvarez and his collaborators used to "x-ray" the pyramid.

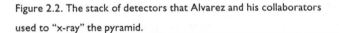

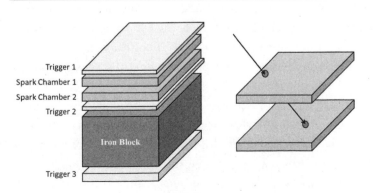

Trigger 1
Spark Chamber 1
Spark Chamber 2
Trigger 2

Iron Block

Trigger 3

While this looks intimidatingly complicated, the key idea is the same as the cosmic-ray telescope Alvarez built as a young student. The three trigger detectors, like the Geiger counters in his telescope, were connected in a coincidence circuit, so they would only record a muon if the particle passed through all three detectors at nearly the same time. The iron slab blocked low-energy muons from being detected, as they might be deflected in passing through the rock.

The big advance in the thirty-odd years between Alvarez's student telescope and the pyramid project was the invention of the spark chamber, a detector that records the exact position of a particle passing through it. When all three trigger detectors recorded the arrival of a muon, the spark chambers would be triggered to measure the coordinates where the muon passed through. Connecting the two spark detector records with a line reveals the path followed by the muon (as shown in the right side of the figure) and thus what part of the pyramid it passed through. Unlike the original telescope, which could measure only along a single line, the spark chambers allowed the simultaneous measurement of muons coming along any line within about thirty-five degrees of the top of the pyramid.

The initial run of the experiment was interrupted by the Six-Day War between Egypt and Israel, but once international tensions eased a bit, in early 1968, the experiment resumed and collected data for several months. The resulting data clearly demonstrated the power of the technique, easily picking out the four corners of the pyramid and details of the limestone cap at the top. Despite the hopes of archeological glory, though, and contrary to expectations regarding Khafre's ego, there was no sign of any hidden chambers. In follow-on experiments, Alvarez and his colleagues rotated the apparatus to cover the parts of the pyramid outside the central cone of the initial measurements, with the same result: The pyramid is solid, all the way through.

This was not merely an inconclusive measurement, though. To paraphrase Alvarez's description slightly, the researchers didn't *fail* to find a hidden chamber in the pyramid; they showed that there *is no hidden chamber to be found*. The method worked and is arguably the only practical use anyone has found for cosmic rays.*

ARTIFICIAL RADIOACTIVITY AND DINOSAURS

The most famous of Alvarez's contributions to other sciences came very late in his career, but like the pyramid project, it can be traced to experiments he did early on. Shortly after completing his Ph.D. with Compton, Alvarez joined the Radiation Laboratory at Berkeley. This lab was at the cutting edge of nuclear physics research and the source of many important discoveries using the cyclotron particle accelerator recently invented by laboratory director Ernest Lawrence.†

One of Alvarez's first independent contributions to the lab was the development of a beam of slow-moving neutrons. The neutron, as its name suggests, is a neutral particle. It has almost the same mass as a proton and is normally found in the nucleus of atoms. Neutrons had been discovered only in 1932, by James Chadwick in Ernest Rutherford's laboratory in the United Kingdom, and research on their properties and reactions was still new and exciting.

---

* Similar projects using cosmic rays to study pyramids in Central America are under way at the ancient Mayan city of La Milpa and at the Pyramid of the Sun at Teotihuacan.

† Lawrence won the 1939 Nobel Prize in Physics for inventing the cyclotron, which revolutionized nuclear physics in the 1930s.

Under the right circumstances, a neutron encountering an atom can be absorbed into the nucleus of that atom, creating a new isotope, a heavier version of the same element. This process depends very much on the energy of the neutrons, and many techniques were tried to produce neutrons of different energies, with little success. Most of the neutron sources available spray out both fast-moving and slow-moving neutrons over a wide range of energies, and separating the two is very difficult.

Alvarez realized, though, that he could make a source of "slow" neutrons simply by placing his detector at the end of a long pipe. As they travel a distance of several meters, the neutrons get strung out in space, like runners in a race: The fast ones reach the end of the pipe quickly, whereas the slow ones straggle in sometime later. By sending a pulse of neutrons into the pipe and only turning on his detector a few milliseconds later, Alvarez could look at only the slow-moving neutrons, with the exact speed determined by the delay between sending in the neutrons and turning on the detector.

In the process, as he describes in his autobiography, he rediscovered *artificial radioactivity*. When certain atoms absorb extra neutrons, the new isotopes produced are unstable and will undergo radioactive decay a short time later, spitting out new particles and changing to a different element.* This process had been discovered in 1934 by Frédéric Joliot and Irène Joliot-Curie in Paris (more on them in Chapter 12), but Alvarez noticed it anew in his slow-neutron experiments: Lithium atoms in the target he was using for his experiments absorbed

---

* Some naturally occurring elements, particularly heavy ones, also undergo radioactive decay, generally with much longer lifetimes than the elements produced in artificial radioactivity. Demonstrating that radioactive decay led to the transmutation of one element into another won Ernest Rutherford his Nobel Prize in Chemistry, as mentioned in Chapter 1.

neutrons and became unstable, leading to large bursts of radio-activity even after the cyclotron was turned off.

Although he wasn't the first to observe artificial radioactivity, the process was to play a key role in the rest of his career. His next major discovery used artificial radioactivity to settle a question about isotopes of hydrogen and helium. At the time, physicists believed that tritium, an isotope of hydrogen with one proton and two neutrons, would be stable, while helium-3, an isotope containing two protons and one neutron, would be unstable. Using the Berkeley cyclotron, Alvarez showed that this was exactly backward, first by detecting helium-3 atoms in samples of helium extracted from natural gas wells (which had spent millions of years deep underground, during which any radioactive atoms would have decayed away), then by creating artificial tritium by bombarding hydrogen with particles from the accelerator and showing that the resulting gas was highly radioactive. The production of radioactive isotopes became a big part of the business of nuclear physics, which continues to this day. Cyclotrons and nuclear reactors are used to produce the radioactive isotopes needed for many medical procedures—radiation therapy for cancer patients and tracer isotopes for medical imaging—and even household smoke detectors, which contain a tiny amount of americium, an artificial element created in nuclear reactors. Radioactive isotope production was also critical to Alvarez's most famous discovery.

The discovery that got Luis Alvarez the most attention from the public came courtesy of his son, Walter, a geologist who was studying reversals of the Earth's magnetic field in Italy. On a visit home, he presented his father with a sample of rock showing the boundary between the Cretaceous era (when dinosaurs were common) and the Tertiary era, which began after the extinction of dinosaurs and many other species.

Walter pointed out the boundary, which was marked by a thin layer of clay between two bands of limestone, and showed his father the dramatic change in the number of fossils across that boundary.

The obvious question to ask is, What made that clay layer? Walter told his father that while the clay layer was found in many places around the world, nobody knew what caused it. This presented an irresistible challenge, so the two Alvarezes set out to find a way to explain the clay, drawing on both Walter's experience with geology and Luis's experience as a physicist.

The first question to answer was how long it took for the clay layer to be deposited. This is a tricky problem, but Luis hit on an unusual idea for a clock. In addition to the cosmic rays that strike the Earth's upper atmosphere, our planet is also constantly sweeping up residual dust left over from the formation of the solar system. The amount of dust is minute, but it does accumulate, and Luis suggested that they could use the dust to estimate the time required to make that clay: The more space dust in the clay, the longer it took to be laid down.

Of course, this calculation would require a way to sort out the space dust from other components of the rock and a way to detect the dust's presence. They realized that the dust could be recognized by its composition, which would be slightly different from the composition of rocks on Earth. Although Earth was formed from the same primordial dust, as the planet cooled and solidified, most of the elements from a group of metals chemically similar to platinum sank into the Earth's core, leaving the rocks of the crust with much lower concentrations of those metals than were found in the original dust cloud. Space dust should contain more of those metals than normal Earth rocks, so the total amount of metals from the platinum group in that clay layer could provide an estimate of how quickly it formed.

Detecting those metals at the tiny abundances that could be expected from the slow accumulation of interplanetary dust presented a formidable technical challenge. But this is where Luis's experience with nuclear physics proved invaluable: He realized that you could detect rare elements in the rock samples by bombarding them with neutrons, to artificially create heavy isotopes. Then, you merely needed to look for the characteristic radiation emitted as those unstable isotopes decayed. One element in particular, iridium, looked especially promising, being more abundant in space dust than Earth rocks. Moreover, iridium had favorable properties for the creation and detection of an artificially radioactive isotope.

Luis enlisted a couple of Berkeley colleagues, chemists Frank Asaro and Helen Michel, to help with the tests, and in 1977, they began irradiating samples of the clay layer and the limestone immediately above and below. In the limestone well above and well below the boundary, they detected the tiny amounts of iridium they expected from accumulating space dust. What they found at the boundary, though, was astonishing: The amount of iridium in the clay was more than thirty times that in the surrounding rocks. Something dramatic had taken place in the formation of that clay. Such a huge amount of iridium couldn't possibly have come from space dust.

After trying and rejecting many explanations, they hit on the answer: A large asteroid, roughly ten kilometers across, had slammed into the Earth at the end of the Cretaceous sixty-five million years ago. In the collision, huge amounts of rock were vaporized and flung into the upper atmosphere, carrying with them the telltale iridium indicating their extraterrestrial origin. The cloud of rock dust quickly spread to cover the entire Earth before settling out of the atmosphere, forming the worldwide clay layer that marks the boundary between the Cretaceous and Tertiary eras.

The impact theory also helps explain the mass extinctions at the end of the Cretaceous. The dust cloud resulting from the impact would have covered the entire Earth for several months, blocking sunlight and killing vegetation. As a result, large, plant-eating dinosaurs would starve to death, followed in short order by the meat-eating dinosaurs that preyed on them. Smaller creatures, such as our tiny mammalian ancestors, could have survived by virtue of requiring less food, though it would have meant lean living for quite a few years.

The Alvarezes published their theory in 1980, and it was controversial for many years—in his 1986 autobiography, Luis Alvarez writes of ongoing arguments with paleontologists favoring other explanations—but supporting evidence for the asteroid impact theory accumulated from several sources. The same pattern of iridium abundance was found at sites all around the world, indicating that the clay was truly worldwide. Microscopic examination of the clay turned up small bits of "shocked" quartz and little rock globules called tektites, indicating a violent origin. And in the early 1990s, an impact site was identified, the Chicxulub crater off the coast of the Yucatan Peninsula, which is the right age, size, and composition to be the source of the clay. These days, the impact theory is widely accepted as at least a factor in the extinction of the dinosaurs—some scientists believe that dinosaurs were already in decline for other reasons and that the asteroid impact was merely the last straw.

BEYOND PHYSICS

The story of Luis Alvarez and his many forays into scientific detective work provides a particularly dramatic example of the power of applying known techniques to new problems in other fields. The general phenomenon is very common in the history of science, though—progress on difficult problems often comes as the result of somebody's bringing new tools to bear.

The double-helix structure of DNA, key to our modern understanding of heredity, was discovered in large part thanks to the work of Rosalind Franklin, a physical chemist working at King's College in London. Franklin had very little training in biochemistry—her previous work had mostly been studying the porosity of different types of coal—but she was an expert on x-ray diffraction, a physics technique using the pattern of x-rays bouncing off a regular structure to deduce the arrangement of atoms and molecules making it up.[*] Franklin used her x-ray apparatus to obtain better diffraction signals than those obtained by anyone else in the world, and one of her images provided the key clue that led to the complete structure of the molecule.[†]

Moving away from physics, the 2002 Nobel Prize in Economics went to Daniel Kahnemann, a cognitive psychologist with no economics background. The prize was well deserved, though, as Kahnemann's work with Amos Tversky on how people make decisions has forced some rethinking of central

---

[*] You can demonstrate the basic idea of x-ray diffraction of DNA using a laser pointer and a ballpoint pen spring: www.t2ah.com/2013/02/number -25-dna-diffraction-with-spring.html.

[†] Franklin's images provided the key data, and she almost certainly would have deduced the structure herself eventually. She was a very cautious and conscientious experimenter, though, and didn't want to rush to conclusions. Her image was shown, without her knowledge, to James Watson and Francis Crick at Cambridge. Watson and Crick realized its importance and ended up being first to publish the correct structure. The complete story can be found in Brenda Maddox, *Rosalind Franklin: The Dark Lady of DNA* (New York: Harper Perennial, 2002).

assumptions of economic theory.* In some beautifully simple psychology experiments, they showed that the rational and self-interested decision making assumed in making simple economic models is an illusion—real people are prone to cognitive shortcuts that introduce significant biases in the choices we make.† The field of behavioral economics launched by Kahnemann and Tversky has proven to be a rich source of new discoveries and economic thinking.

The usefulness of outside expertise carries over into fields outside science, as well. Since the days when we used to watch *Iron Chef* back in New Haven, Kate has provided invaluable assistance on everything I've written—job and grant applications, research papers, and my popular audience books (including this one). This is not because she has any deep knowledge of physics—far from it. As an attorney, though, she has a great deal of experience with persuasive writing and can offer helpful tips on tightening up my arguments and fixing my more egregious abuses of grammar.

So, there is a lesson to take from the examples of Luis Alvarez and the Iron Chefs. When you are confronting a complex problem, it can often be useful to seek input not only from people close to the situation, but also from those well outside the field. A different perspective and a different set of skills and approaches can sometimes provide new insights and enable amazing results.

Another, just as important, bit of advice is this: Know what you do well, and don't be afraid to apply old approaches to new

---

* Tversky died in 1996; otherwise, he probably would have shared the Nobel with Kahnemann.

† These are described in detail in Kahnemann's best-selling book, *Thinking, Fast and Slow* (New York: Farrar, Straus and Giroux, 2011).

problems. This shouldn't be taken too far—you don't want to be the hammer owner who sees everything as a nail. Provided you have a reasonably diverse set of analytical tools, though, translating a new problem into terms you're familiar with and approaching it from that direction can prove fruitful.

IRON CHEF NISKAYUNA

While it can be a challenge to find the time, I generally enjoy cooking and handle most of the food preparation in our house. Nobody is ever going to invite me to Kitchen Stadium to challenge an Iron Chef, but I do all right. And one of my favorite comfort-food recipes can be traced, in part, to the show.

I've always been fond of Americanized Chinese food, and one of my go-to dishes for using leftovers has always been fried rice. I'm not a huge fan of eggs, though, so the traditional method of scrambling an egg and then adding rice to the pan has never been entirely satisfying. On one episode of Iron Chef, a challenger suggested that the secret to good fried rice was to mix the egg with the rice before cooking, which gets much more even distribution. That tweak to the basic technique was a great addition.

It wasn't the end of the evolution of my comfort food, though, as a few months later, I had already dumped the rice in the pan when I realized I was out of soy sauce. Looking for something to flavor the rice, I grabbed a jar of sambal oelek chili paste, and tossed a spoonful of that in. The resulting flavor was good, but a bit more Thai than Chinese, so the next time, I added a bunch of basil, as well, to good effect.

The result is something very different from what I started out making, but quite tasty and a regular part of my comfort-food rotation. It came about through a natural evolution, based on adapting basic techniques and flavor combinations to the ingredients at hand. And while I didn't realize it until I started writing this book, it's also based in an essentially scientific process: looking at what I had on hand, thinking of what might work well, testing that model, and refining the process for the next batch.

I'm not suggesting that my method is a revolutionary breakthrough in cookery or in anything else—quite the contrary. The example is a simple illustration of the sort of thing that any decent cook should be able to do when faced with new or missing ingredients. And if you have enough confidence in the kitchen to adapt known methods to turn a pile of miscellaneous ingredients into a meal, you should be able to do the same and bring familiar mental skills to bear on new problems. You can also rest assured that this approach has a long and distinguished history of producing scientific breakthroughs.

Chapter 3

NEEDLES IN HAYSTACKS

We live a couple of miles from Union College's Schenectady campus, in an affluent, older suburban neighborhood—most of the houses date from the World War II era (ours was built in 1941). It's a great neighborhood for walking, with quiet, tree-lined streets, and every morning, the dog and I go out for a stroll around the neighborhood.

On our Saturday morning walks in the summer months, we usually pass at least one yard sale, sometimes more.* A lot of the market for these is, as you would expect, students and young families with relatively little money. The sales also draw a substantial number of people from outside the neighborhood, though, who don't need cheap dorm-room furniture or secondhand baby goods. These folks are generally engaged in a particular form of speculation: They're hoping to find valuable items at a steep discount, to either add to a collection or sell to other collectors at a profit. Even though our morning dog walks are before the official start time for most of these

---

* Around the corner from us is one house whose owners I suspect of selling other people's stuff on consignment, because they seem to have a yard sale every weekend.

sales, we'll almost always see a few cars cruising the sales, trying to get a jump on the competition.

This kind of yard-sale speculation is related to two popular television genres. The first is exemplified by the show *Antiques Roadshow*, which Kate sometimes watches for relaxation. The show features people bringing odds and ends—family heirlooms, old stuff found in attics, and the occasional yard-sale purchase—to be assessed by experts on various types of antiques. Every episode features at least one family discovering that some trinket—the odd little statue Great-Aunt Sally had sitting in her parlor—is worth tens of thousands of dollars. A related genre comes at this from the perspective of the buyers—*American Pickers, Pawn Stars,* and other shows featuring dealers who try to obtain valuable merchandise cheaply from people who aren't aware of what they have.

While many people entertain fantasies of stumbling across some incredibly valuable antique at a yard sale, the TV versions undersell the difficulty of the enterprise. The vast majority of yard-sale merchandise is basically worthless.* The people delighted to discover they've been sitting on priceless art on *Antiques Roadshow* are selected from a huge pool of people who are sent home disappointed. Many of these "nonwinners" are visible in the background—the shows are filmed inside convention centers full of people with high hopes for the old crap they found in their attic.

Still, there's no shortage of people who cruise around yard sales hoping to stumble across a valuable needle in the secondhand haystack. Success in this sort of speculation requires a good deal of ingenuity—finding unusual places to

---

* From an investment standpoint, at least—if you're a student in need of a cheap futon, it can be priceless.

look, knowing shortcuts for quickly spotting things that might be worth purchasing, and knowing how to haggle the sellers down to a lower price to maximize the return. The biggest returns probably come from spotting items that are in poor condition at the time of sale, but might be sold at a significant profit if properly restored. Mostly, though, it requires the patience to look through lots and lots of junk in hopes of finding a tiny number of valuable antiques.

This kind of patient searching for exceedingly rare events is also a big part of science. Many areas of science require sifting through a vast mountain of crap to unearth a tiny number of interesting items. Paleontologists dig through huge piles of fossil fragments hoping to turn up a complete skeleton; field biologists peer at hundreds of undistinguished beetles hoping to turn up a new species; comet hunters peer at vast numbers of stars, night after night, looking for one that changes position. Whole scientific careers are built on the hope of tracking down some elusive event.[†]

Perhaps the most elusive prey successfully tracked down by scientists, though, is the neutrino. These subatomic particles are created in vast numbers in the sun and other stars, but neutrinos interact with everyday matter so weakly that we don't even notice—trillions of neutrinos will pass through you as you read this sentence, with no noticeable effect. Thanks to heroic efforts by many scientists over a span of decades, though, neutrinos have been detected, and their behavior

---

[†] This quest sometimes tips over into pseudoscience, as in the pursuit of many "cryptozoological" creatures like Bigfoot and the Loch Ness Monster. The key to keeping the search scientific is to avoid letting hope overwhelm analysis—too many cryptozoologists let their faith in the existence of their quarry make them too credulous when confronting poor-quality eyewitness testimony and the occasional active hoaxer.

provides some of the best hints we have of physics beyond the Standard Model. The hunt for these elusive particles is a great illustration of how scientists can benefit from the persistence and patience characteristic of antique hunters.

"I HAVE DONE A TERRIBLE THING"

While neutrinos are fundamental particles and the universe still contains huge numbers of neutrinos created during the Big Bang, because they interact so weakly with ordinary matter, their existence is not immediately obvious.* For that reason, their history in modern physics begins as an act of theoretical desperation.

As quantum mechanics was being worked out in the 1920s, physicists found explanations for all sorts of previously mysterious processes. When the structure of atoms was nailed down, the frontier of understanding moved inward, to the nucleus of the atom. Issues of nuclear physics proved more difficult, however, particularly when it came to radioactivity, the tendency of certain types of atoms to spit out high-energy particles and change their nature.

Around 1900, Ernest Rutherford introduced the Greek letter system still used to classify the decay of unstable elements by the type of radiation emitted. In alpha decay, a heavy nucleus spits out a helium nucleus (an *alpha particle*). In beta decay, the decaying nucleus spits out an electron or a positron (these are called negative or positive beta particles in this context, a bit of terminology that occasionally creates confusion). And in gamma decay (not part of Rutherford's original

---

* Several hundred of these relic neutrinos are thought to exist in every cubic centimeter of the universe.

scheme, but added soon after its identification by Paul Villard) the nucleus spits out a high-energy photon. Both alpha and beta decay change the nature of the nucleus, and Rutherford won the 1908 Nobel Prize in Chemistry for demonstrating this transmutation of elements.

As physicists began studying nuclei in detail, alpha particles, being helium nuclei themselves, were readily understood as natural components of the nuclei of atoms. Alpha decay represented the splitting apart of a too-heavy nucleus. Gamma decay doesn't involve transmutation and can be understood as a rearrangement of the particles within the nucleus.[†] Beta radiation, however, proved very difficult to understand.

In the late 1920s, the only subatomic particles known were the proton and the electron, so beta radiation was initially thought to be the escape of an electron that was somehow trapped within the nucleus. That model predicts the emission of beta particles of a single, specific energy—the total amount of energy to be released in the decay is split between the daughter nucleus and the emitted electron in proportion to their masses (the electron heading in one direction at high speed and the vastly more massive nucleus recoiling in the opposite direction at a lower speed).[‡] In reality, however, beta decay produces electrons of a whole range of energies, up to some maximum value. There's no way to explain this spread of energies using only two particles.

[†] Gamma decay often follows immediately after an alpha or beta decay that does involve a change of element, though, as the new nucleus is left in an unstable configuration, and then rearranges itself by emitting a gamma ray.

[‡] Alpha decay works exactly this way—the emitted alpha particles have a very narrow range of energy, and Rutherford made use of this for the experiments that discovered the existence of the nucleus in 1909 (Chapter 8).

In 1930, the famously sardonic Austrian physicist Wolfgang Pauli, in a letter addressed to the "Dear Radioactive Ladies and Gentlemen" at a conference he was unable to attend, proposed a radical solution to the problem: Rather than the two particles believed to be involved, beta decay must involve a third, unseen particle. The total energy released in beta decay is always the same, but some part of it is carried away by Pauli's new particle. Pauli originally called his particle a "neutron," but that name was grabbed by a heavy neutral particle, the third component of ordinary atoms, discovered by James Chadwick in 1932. The name we use today—neutrino—was applied by the Italian physicist Enrico Fermi. With the diminutive ending -ino, the term translates to something like "little neutral one."

Pauli was initially ambivalent about his proposal, as he explained in a letter to Walter Baade: "I have done something terrible. I have postulated a particle that cannot be detected. That is something a theorist should never do."* He regarded the alternatives as much worse, however—Niels Bohr had proposed revoking the idea of energy conservation as a fundamental principle of physics—so he stuck with his undetectable particle. Fermi's assistance put the neutrino on a more solid footing, and today, physicists understand beta decay as the disintegration of a neutron in the nucleus, transforming into a proton, an electron, and a neutrino.† Although some prominent physicists remained uncomfortable with the idea of the

---

* Quoted in Ray Jayawardhana, *Neutrino Hunters: The Thrilling Chase for a Ghostly Particle to Unlock the Secrets of the Universe* (New York: Scientific American/Farrar, Straus and Giroux, 2013), 42.

† This takes place through a new interaction, the *weak nuclear force*, so when physicists say neutrinos interact only weakly with other matter, they mean it literally. The theory of the weak force was confirmed by the detection of the W and Z particles in the 1980s.

neutrino, the practical success of the theory was undeniable, correctly predicting most of the properties of beta decay.

The lack of a direct observation of the particle continued to bother physicists, though, and people sought a way to detect the "undetectable" neutrino. One glimmer of hope came from the possibility of an inverse beta-decay process. If neutrinos are created through the radioactive decay of ordinary matter, the essential symmetry of physics demands that they must be able to interact with ordinary matter and trigger other decays. The exact reversal of beta decay is vanishingly unlikely—it would require a proton, an electron, and a neutrino to come together at the same instant—but another process is possible: a proton could absorb a neutrino and split into a neutron and a positron. Hans Bethe and Rudolf Peierls calculated the likelihood of this process in 1936 but decided that "it seems practically impossible to detect neutrinos"—the probability of seeing inverse beta decay was just too low, given the neutrino sources known at the time.

DETECTING THE UNDETECTABLE

While Bethe and Peierls were willing to (relatively) confidently declare that neutrinos would not be detected, another prominent neutrino doubter was more cautious. Arthur Eddington, a British astrophysicist and popular author famous for an expedition that confirmed Einstein's general theory of relativity, reframed the question this way: "Dare I say that experimental physicists will not have sufficient ingenuity to make neutrinos?" Although he was skeptical about the theory, he was not confident that practical issues were enough to keep them from being detected: "I am not going to be lured into a wager against the skill of experimenters under the impression that it is a wager against the truth of a theory." This caution would prove wise, as by 1956, neutrinos had, in fact, been definitively detected.

The 1956 experiment was the work of two physicists at Los Alamos, Frederick Reines and Clyde Cowan Jr. They realized that while the probability of inverse beta decay was extremely low, given a big enough detector and an intense enough source of neutrinos, it should still be possible to directly detect neutrinos. And thanks to the Manhattan Project during World War II, Reines and Cowan had access to extremely intense sources of neutrinos. Natural radioactivity doesn't produce enough neutrinos to be easily detected, but nuclear reactors and nuclear weapons produce neutrinos in vastly greater numbers. Reines and Cowan decided to use a large tank containing hundreds of liters of a liquid scintillator material—a substance that creates small flashes of visible light when interacting with radioactive particles—and look for the traces of inverse beta decay.

Even with several hundred liters of scintillator, though, the physicists would have needed a truly gigantic number of neutrinos to detect anything. The problem is that natural radioactivity and cosmic rays also produce particles that can make flashes of light in their scintillator. Like yard sale shoppers looking for antiques, the researchers would be confronted with a vast amount of worthless background hiding their valuable neutrino signal.

For the number of detected neutrinos to rise above this natural background, Reines and Cowan initially thought they would need the kind of neutrino production only found in a nuclear weapons test. Indeed, that's what they proposed to do, via a complicated scheme: Their detector tank would be dropped at the time of detonation of a nuclear bomb during a test, fall into a deep hole, and land on a cushion of feathers and rubber. Reines and Cowan would retrieve the detector and check their results several days later, once the radioactivity from the test had dropped to a safe level. While they deserve full credit for ingenuity and ambition, they were saved from such extreme measures by some ingenuity: a clever scheme to separate the small number of detected neutrinos from the background of other radiation.

Their initial plan was to detect the scintillation light produced when the positrons emitted in the inverse beta decay reaction were annihilated by the electrons in the scintillator, producing gamma rays. They realized, however, that with a small adjustment, they could also detect the neutrons produced in the reaction. These neutrons escape from the nuclei of the original atoms and would bounce around the tank, colliding with molecules in the liquid. Within a few microseconds, the neutrons would lose enough energy to be absorbed by a small amount of cadmium added to the scintillation liquid, creating a second flash of light. By looking not for a single scintillation flash, but *two* flashes within a few microseconds— each flash within a particular range of energies—the physicists could discriminate between neutrinos and other types of background radiation.[*] This greater discrimination would allow Reines and Cowan to avoid the nuclear test range and instead make do with the much smaller number of neutrinos produced by the operation of a nuclear reactor.

An initial test in 1953 with a 300-liter detector at the Hanford nuclear reactor in Washington State showed clear hints of neutrinos—the rate of particle detections was higher when

---

[*] This is the same sort of coincidence counting used for Luis Alvarez's cosmic-ray telescopes in Chapter 2.

the reactor was running—but did not provide conclusive evidence. Reines and Cowan's second experiment, in 1956, used a larger detector (400 liters of water and 4,200 liters of scintillation liquid) and a more powerful reactor, the Savannah River plant in South Carolina. After weeks of operation—some 900 hours with the reactor running and 250 hours with the reactor off—they had their conclusive proof. Even with all their efforts to maximize the detection efficiency, the rate was tiny—they detected about one neutrino per hour from the reactor—but the signal was five times larger when the reactor was on than when it was off, and numerous consistency checks confirmed that the signal was due to neutrinos. Reines and Cowan sent a congratulatory telegram to Wolfgang Pauli, who interrupted a meeting he was attending to read their announcement to the other physicists present, then celebrated by drinking a case of champagne with friends.* Twenty-six years after Pauli's initial proposal, his terrible, undetectable particle had finally been detected.

UNDERGROUND ASTRONOMY

The detection of neutrinos finally clinched the status of the theory of nuclear decay worked out by Pauli and Fermi, and it might seem that this discovery would be the end of the subject. Over the last fifty years, though, many millions of dollars have been spent building bigger and better neutrino detectors. Neutrinos, it turns out, are not just a theoretical curiosity, but provide an important window for studying the larger universe.

Neutrinos are produced not only in nuclear reactors on Earth, but in space as well. In particular, the nuclear fusion reactions that power the Sun produce vast numbers of neutrinos.

---

* All figures and anecdotes from "The Reines-Cowan Experiments: Detecting the Poltergeist," *Los Alamos Science* 25 (1997).

The Sun shines by fusing the nuclei of hydrogen atoms (which consist of a single proton) into helium nuclei (consisting of two protons and one or two neutrons)—along the way, some of the original protons must be turned into neutrons, emitting a neutrino in the process. Neutrinos produced in the core of the Sun pass effortlessly through the outer layers of the Sun, while a photon of light produced in the same spot takes hundreds of thousands of years to make its way out, because of constant interactions with the solar atmosphere. Thus, neutrinos from the Sun are one of the best tools we have for studying the reactions in the core of the Sun.

These neutrinos are produced in vast numbers—tens of billions of them pass through every square centimeter of the Earth every second—but are extremely difficult to detect. Gigantic detectors are needed to pick up tiny numbers of neutrinos. In the decades since Reines and Cowan made their measurement, two major techniques have been used for neutrino detection. One is a straightforward extension of the scintillation method: the detection of tiny flashes of light produced by neutrinos interacting with ordinary matter. Several variations on this have been performed over the years, and Masatoshi Koshiba shared the 2002 Nobel Prize in Physics for developing one of the most important scintillation-based neutrino detectors, the Kamiokande experiment in Japan. This technique uses a huge volume of water as the scintillation medium—50 million liters in the current Super-Kamiokande configuration—surrounded by thousands of photodetectors to pick up the light flashes. The instruments distinguish neutrinos from other particles by the characteristic pattern of radiation created when a neutrino interacts with a particle in the water.

One of Koshiba's colaureates, however, Raymond Davis Jr., won for his work on what may be the ultimate needle-in-a-haystack experiment, detecting neutrinos through the by-products of the inverse beta decay used by Reines and Cowan.* Inverse beta decay changes protons to neutrons, which in turn changes the nature of the atoms containing those protons. Davis realized that the reaction could turn atoms of chlorine into a radioactive isotope of argon. Rather than trying to distinguish scintillation flashes caused by neutrinos from those caused by other radiation, Davis decided to look for the argon directly.

Of course, since neutrinos are so weakly interacting, to have any hope of detecting them requires a very large amount of chlorine, so Davis assembled a 380,000-liter tank and filled it with tanker trucks worth of dry-cleaning fluid. Even with such a large target, though, the number of argon atoms produced is tiny—a few dozen a month, give or take. This might seem an impossible task, but Davis proved he could do it by deliberately introducing five hundred atoms of argon to the tank and showing that he could detect every one.†

Davis's experiment operated continuously from 1970 to 1994, sifting a handful of neutrino-produced argon atoms out of the tank every few months. This showed that there were, indeed, neutrinos produced by the Sun, but about one-third the number expected from theoretical models. When the Kamiokande detector began working in the 1980s, it confirmed Davis's value for the total number, deepening the mystery. Physicists eventually determined that neutrinos come in multiple "flavors," only one of which was being detected. The "missing" neutrinos from the Sun were changing from one type to another during their eight-minute flight from the Sun to the

---

* The third laureate that year, Riccardo Giacconi, won for work in x-ray astronomy, which is a fascinating subject in its own right, but not really relevant here.

† He modestly characterized the process of finding the argon atoms as "just plumbing."

Earth. The Kamiokande experiment verified this *neutrino oscillation* through a variation in the number detected at different distances from the Sun, and the results have since been nailed down by the Sudbury Neutrino Observatory in western Ontario and recent experiments at Kamiokande and Daya Bay in China with neutrinos produced by nearby reactors.

Neutrino oscillation is a major discovery because it means that contrary to Pauli's initial expectation, neutrinos must have mass. The masses are tiny—around a millionth of the mass of the electron—but definitely not zero. Neutrino masses were not accounted for in the Standard Model of particle physics, so this discovery is one of the best hints we have of physics beyond what's well known.

Neutrino astronomy is not limited to probing reactions in the Sun. Some 158,000 years ago, a giant star in the Large Magellanic Cloud exploded; the light from that explosion reached Earth in 1987, recorded under the prosaic name Supernova 1987A. Along with that light came a gigantic burst of neutrinos—accounting for the vast majority of the energy released in the explosion—and about two dozen of these were detected at neutrino laboratories around the world (eleven in Kamiokande, eight at the IMB [Irvine-Michigan-Brookhaven] detector in the United States, and five at the Baksan neutrino observatory in the Caucasus Mountains). These represent the first unambiguous evidence of neutrinos from beyond the Solar System, and hundreds of papers analyzing that tiny cluster of particles and using them to test theories of stellar explosions have been written over the past twenty-five years. Neutrino detectors have improved substantially in a quarter century, so neutrino hunters and astrophysicists are eagerly awaiting the next nearby supernova, which should provide a far greater wealth of scientific data.

As impressively scanty as the number of detected neutrinos is, even more impressive is the lengths neutrino hunters must go to just to get those few particles. Neutrino observatories are some of the most important sites for studying the universe, but all of them are located in places where you can't see the sky. Ray Davis's tank of cleaning fluid was at the bottom of the Homestake mine in South Dakota, Super-Kamiokande is in the Mozumi mine near Hida, Japan, and the Sudbury Neutrino Observatory is more than a mile deep in the Creighton nickel mine in Ontario.

Deep mines may seem an unusual venue for astrophysics, but the location is essential to their success. While neutrinos outnumber all other forms of radiation, they interact so weakly that less numerous cosmic ray particles are vastly more likely to be detected. Screening out enough of the cosmic rays that bombard the Earth from space to allow neutrinos to be detected with confidence requires that the detectors be buried under hundreds of meters of solid rock, or even more. The newest and largest neutrino observatory in the world, the IceCube South Pole Neutrino Observatory in Antarctica, uses a cubic kilometer of ice as a scintillation medium. But even though the IceCube detectors are buried a mile deep, they only count particles headed *up*—that is, particles that passed through the entire Earth before reaching the detectors. Even a mile down, the majority of the downward-traveling particles come from cosmic rays. The entire mass of the planet isn't enough to deter the passage of neutrinos from the Sun and other cosmic sources.

In November 2013 IceCube published a paper on twenty-eight extremely high-energy neutrinos—energies a million times greater than those from Supernova 1987A and at least ten times greater than any particle accelerator on Earth—collected over two years of operation. These particles originate outside the solar system, probably in astrophysical accelerators like black holes, and were only detected in significant numbers because of the huge scale of the detector and its careful screening of background noise. Like antique hunters picking through piles of junk at a yard sale and coming up with a priceless work of art, these scientists saw their patience and perseverance pay off and may be ushering in the start of a new era of astronomy, deep underground.

OTHER SCIENCES

Neutrino hunters have far and away the lowest success rate in science, in terms of the number of particles they see compared with the number of particles they're looking for—the two dozen neutrinos detected from Supernova 1987A are out of an estimated 10^{58} produced in the explosion, or about ten neutrinos for every proton in the Sun. All sciences benefit from patience and persistence, though.

Marie Skłodowska Curie is one of only four people to win two Nobel Prizes, sharing the 1903 prize in physics with her husband Pierre and Henri Becquerel for studies of radioactivity, and winning the 1911 prize by herself for the discovery of the element radium.* Becquerel was the first scientist to notice the phenomenon of radioactivity, in 1896, when he found that a jar of uranium salts exposed a photographic plate stored nearby. A short time later, Marie noticed that pitchblende, the ore from which uranium is refined, is more radioactive than can be explained by

---

* She undoubtedly would have shared the second Nobel with Pierre, but he was killed in a road accident in 1906, and Nobel Prizes are never awarded posthumously.

the uranium alone, and she set out to isolate the substances responsible for the extra radioactivity. Marie and Pierre identified two additional radioactive substances in pitchblende ore, both more radioactive than uranium. They named one *polonium* after the nation of Marie's birth and the other *radium*, though they were only able to produce tiny amounts of these substances. To study the properties of the new elements more effectively, they needed to work with large quantities of material.

Pitchblende was very expensive, more than the meager budget available to the Curies, but they found a mine in Bohemia that was happy to donate a large amount of the residue left over after the uranium was extracted. Beginning with a metric ton of pitchblende residue, Marie and Pierre worked for three years to isolate a tenth of a gram of radium chloride in 1902. Separating radium was an arduous process involving dozens of steps, including such fun activities as boiling the ore in caustic soda and washing it in concentrated hydrochloric acid. All of this was conducted in a "miserable old shed" in the courtyard at the École Normale Supérieure in Paris.* The German chemist Wilhelm Ostwald described the facility as "a cross between a stable and a potato cellar." Despite the conditions, Marie Curie described this period as the "best and happiest years of our life . . . entirely consecrated to work."† Still more

---

* The Curies were never well funded and refused to patent their discoveries and thus profit from them. Despite discovering radium, Marie was never able to afford large quantities of it to study; this ironic situation caught the attention of the American journalist Marie Meloney, who launched a campaign to collect donations from the women of America to raise the $100,000 needed to purchase a gram of radium and donate it to Curie's lab. The gift was presented by President Warren Harding in 1921.

† These years also wrecked their health. The hazards of working with radioactivity were, of course, not yet known, and both Pierre and Marie showed symptoms of severe radiation sickness from long stretches of working and even eating their meals in the shed.

hard chemical labor eventually let Marie produce pure radium metal in 1910; she was never able to isolate pure polonium, as it decays too quickly. For her patience and persistence alone, Curie's two Nobels rank among the more richly deserved in the history of the prizes.

Experiments stretching across decades aren't limited to physics and chemistry. Zoologists frequently devote years to studying the behavior of animals in the wild, and primatologists like Jane Goodall and Dian Fossey became internationally known for spending years at a time observing and interacting with chimpanzees and gorillas, respectively, in remote areas in Africa. And in astronomy and planetary science, large experiments often take many years to construct, particularly when they need to be launched into space. The New Horizons probe headed for Pluto was years in development before being launched in 2006 and won't even reach its destination until 2015.

The experiment demanding the most patience of all, though, may be the pitch drop experiment at the University of Queensland. The experiment started with a mass of hot pitch poured into a sealed funnel in 1927; after the mass was allowed to cool and settle for three years, the end of the funnel was snipped open in 1930. Once cooled, pitch is hard and brittle, but still a liquid, and it flows very slowly through the funnel, dripping into a beaker at a rate of about one drop a decade. The final separation of the drops from the end of the funnel happens very quickly and has never been seen in real time; the most recent custodian of the experiment, John Mainstone, who took charge of it in 1961, died in 2013 without ever seeing a drop fall.‡

---

‡ Five drops fell during Mainstone's watch, but none were witnessed. The eighth drop since 1930 fell in 2000, but the camera set up to record the fall malfunctioned at the critical instant. A similar experiment in Dublin, begun in 1944, successfully recorded video of a falling drop in 2013.

BOOK HUNTING

I've never gotten into antiques, but I do read a lot, particularly science fiction and fantasy novels. I also spent a long time in graduate school, with very little money, so I got in the habit of frequenting used book stores, looking for cheap copies of the books I wanted. Kate shares much of the same taste in books, so until our recent switch to mostly reading electronic books, we used to make these bookstores a regular part of our shopping trips.

While we could usually find specific books of interest at specialty stores devoted to genre fiction, the owners of these stores tended to know the actual value of their merchandise, and the stuff we wanted was usually priced out of our range. Stores selling general-interest used books were of far more interest, as they would often have paperback copies of out-of-print novels available at a steep discount.* We found multiple copies of one of our favorite out-of-print titles on the twenty-five-cent rack at one local store and used to ship them out to friends. But probably our best used-book find was a copy of Glen Cook's *The Dragon Never Sleeps*, which we stumbled across on the dollar rack in an antique store in Springfield, Missouri, while waiting for a friend's wedding. The book had a small print-run from a publisher that went out of business shortly after the book's release, and had been out of print for fifteen years at that point. Sadly, that copy disappeared from my checked bag on the flight home—apparently there are space opera fans at the TSA—and a replacement copy ordered over the Internet ran almost twenty dollars.

---

* These generally weren't in good enough condition for the collector's market, but we wanted them to read, not as an investment, so that wasn't a problem.

That's not exactly a story to anchor an episode of *Antiques Roadshow*, but it does demonstrate the universality of patient searching. Anyone with a serious hobby has undoubtedly experienced this at one time or another: hours of searching in obscure places turning up a crucial missing item for a collection or the raw material for a new project. That kind of patience and perseverance is essential for science. If you've ever found an antique at a yard sale or an obscure book on the dollar rack, you know a tiny bit of what it's like to be a neutrino hunter.

Chapter 4

WALDO AT THE GALAXY ZOO

Anybody who was around in the mid-1980s probably remembers the introduction of the *Where's Waldo?* series of books, which became a mass cultural phenomenon in 1987.* The books feature a series of two-page spreads drawn by illustrator Martin Handford, in which the title character in his signature red-and-white striped shirt is hidden among dozens of other figures in complicated comic scenes. The goal is to find Waldo and several other recurring characters in each of the scenes. The task gets more and more difficult as the books go along, with Waldo only partly visible in the later volumes (there are at least six) and other characters in red and white crowding the scene.

Where's Waldo? is the most famous example of a large genre of find-the-hidden-object books and games aimed at children and some adults—the numerous *I Spy* books by Walter Wick and Jean Marzollo, using photographs of large collections of toys and other small objects, are another staple of

---

* These were originally published in the United Kingdom as *Where's Wally?*, but for unfathomable reasons, the name was changed for the US edition, which is the one I know.

children's bookstores. Our five-year-old daughter is a big fan of both *Where's Waldo?* and *I Spy* and is often better at locating the target objects than I am. Her two-year-old brother doesn't quite have the idea of finding Waldo yet, but finds the pictures fascinating and delights in pointing out larger objects.

While such books may seem like merely an amusing diversion for children, the mental process involved in finding Waldo and his friends in Handford's elaborate drawings is remarkably sophisticated. There are multiple websites and academic papers devoted to computer algorithms for locating Waldo within Handford's drawings, using a variety of software packages that are impressively complex, running to hundreds of lines of code and invoking sophisticated image-processing tools. Child's play this is not.

The essential element of these books is pattern matching, looking for a particular arrangement of colors and shapes in the midst of a distracting field. There are numerous more "adult" variations on this game, some of them obvious, like the image-based hidden-object puzzles Kate sometimes plays for relaxation or the classic video game Myst. Other classes of games may not seem directly connected, but use the same pattern-finding tricks, such as solitaire card games like Free Cell (my own go-to time-waster) or colored-blob-matching games like the massively popular Candy Crush. In all of these, the key to the game is spotting a useful pattern within a large collection of visual data. This is a task at which human brains excel, and millions of people do it for fun and relaxation.

The pattern-matching process involved in finding particular figures within complex scenes is a frequent component of scientific research and a talent much in demand in some fields. Astronomy in particular has benefited greatly from people's ability to spot unusual patterns. Pattern matching is also the

key to one of the largest collections of "citizen-science" projects, providing an opportunity for interested nonscientists to
contribute to scientific progress from the comfort of their own
web browsers.

HUMANS VERSUS COMPUTERS

In the early days of computing, as people began to talk about
artificial intelligence, many people suggested games like chess
as a sort of gold standard for judging the success of a machine's
intelligence. The very best human chess players are essentially
sorting through an incredibly complicated series of possible
moves, working out the possible consequences of each of their
own moves, and predicting what their opponent might do in
response. It's something only a tiny number of humans can
do at the highest levels, after years of study and practice, so it
seemed obvious that playing chess expertly would be nearly
impossible for a computer. Computers might be adept at
crunching numbers, people thought, but sorting through the
possibilities of a chess game to plan and execute a strategy was
something that only humans could manage.

This belief was dramatically shown to be wrong in 1997,
when Deep Blue, a chess-playing computer designed by IBM,
defeated human grand master Gary Kasparov in a series of six
games—the first time a computer program had defeated a human champion. Several subsequent matches have shown that
chess programs running on commercially available computers
can beat all but the very best human players, and humans are no
longer a match for computers specifically designed to play chess.

As is often the case, the triumph of chess-playing computers seems obvious in retrospect. Although the computational
problem presented by chess is difficult, it's also limited: There

are only a finite number of possible moves at any stage of the game, and a sufficiently powerful computer, with proper programming, can sort through them more efficiently than a human brain. The exponential increase of computing power over the last several decades made a computer chess champion inevitable—eventually, computers would be fast enough to defeat humans through sheer brute force, if nothing else.

Of course, the dethroning of humans as chess champions did not usher in a new era of artificial intelligence. The goal of a thinking machine is something of a moving target—every time that computers surpass humans at some task, a new benchmark is found that will be the *real* hallmark of artificial intelligence. A colleague in computer science observed that twenty-five years ago, chess playing would have been covered in a course on artificial intelligence, but these days, it's considered algorithm design.

So, what problems are hard for computers these days? One of the biggest is something that we do so effortlessly that it didn't even occur to the pioneers of artificial intelligence to count it as a problem for computers to solve: processing visual input.

We don't think about vision as a difficult problem, but in fact it's an amazing computational feat. The input that our eyes receive is just a two-dimensional map of the intensity of light of various colors. But somehow, we are able to stitch this together into a three-dimensional picture of the world around us. We can identify and track moving objects, distinguish subtle gradations of color and texture, and combine the very slightly different images from our two eyes to make a good estimate of the distance to some object in our field of view.

No step in this process is simple. Just getting a computer to reliably identify the edges of an object turns out to require a surprising amount of computing power, let alone matching up two images to get a three-dimensional view. Tracking an object that's moving and that may consequently be changing its apparent shape is another huge challenge. Even more impressive is our ability to spot and "read" faces, which kicks in at a very early age—infants only a few months old will visually lock on to real and cartoon faces.* Toddlers can outperform even the best computer facial-recognition routines. That we can do all this without apparent effort is a testament to the remarkable power of the human brain.†

This unmatched ability of humans to spot meaningful patterns in visual data is the basis for many scientific discoveries in all sorts of fields. Probably no field has benefited more from pattern matching than astronomy, though; many of the field's most important and unusual discoveries began with the spotting of an odd pattern.

VARIABLE STARS AND DISTANCES

Every human civilization we're aware of has studied astronomy at some level, noticing patterns in the motion of the various lights in the sky. The simplest astronomical observations—of the rising and setting points of the Sun throughout

---

* The ability to spot and read faces sometimes runs amok, leading to the phenomenon of *pareidolia,* where people "see" a face in random patterns of light and shadow, say, in photographs of Mars or in the irregular browning of a piece of toast.

† This is not to suggest that human brains are more powerful processors than computers—in fact, our brains accomplish these feats of visual processing using a number of built-in shortcuts that cognitive scientists are beginning to understand and computer scientists will eventually employ in making machines that "see." These shortcuts are what lead to the many fascinating optical illusions involving figures that appear to move or change size or color—see, for example, Michael Bach, "Optical Illusions & Visual Phenomena," Web page, last updated June 8, 2014, www.michaelbach.de/ot/.

the year—have been recorded in monumental architecture on several continents. The motion of the planets is a bit more complicated, with patterns that require several years to repeat, but again, civilizations in Asia, the Mediterranean, and South America all tracked and predicted the motion of planets over long stretches of time. As the tools used to observe the sky became more sophisticated, astronomers were able to observe smaller and more subtle changes. Rather than tracking the motion of objects across the whole sky, they began to focus on the behavior of smaller pieces of the sky and on changes in single stars and small groups of stars.

From the late 1800s through most of the 1900s, the primary tool for studying astronomy was the glass photographic plate. Astronomers would point their telescopes at a particular region of the sky night after night and record images of all the light coming from that area. These plates could accumulate light for hours at a time, picking up objects far too faint to be seen by the human eye and revealing a vast range of new types of celestial objects for astronomers to study and classify. With this new technology, numerous surveys of the sky were launched. Astronomers collected many photographic images covering the entire sky and repeated the process many times to look for changes in the arrangement or brightness of the various objects detected.

Of course, vast numbers of faint objects showed up on these plates, and sorting out which of the objects were actually interesting was a difficult task, requiring careful study of dozens of images of tiny dots. This tedious and low-status work often fell to poorly paid or even volunteer women working as "computers" at the world's great observatories. These women would spend long hours poring over astronomical images, noting important features to serve as data for the better-paid male astronomers who ran the observatories and did higher-level analysis. Several of these women made groundbreaking contributions to astronomy, particularly Annie Jump Cannon, who over her career mapped and classified more than three hundred thousand stars and developed the stellar classification scheme still used today.

One of women employed at the Harvard Observatory, Henrietta Swan Leavitt, had discovered astronomy as a student at Radcliffe College (the women's college associated with Harvard) in the early 1890s. A severe illness shortly after graduation left her profoundly deaf.* But when she recovered, she volunteered at the Harvard Observatory, then was hired full-time in 1902 for the wage of twenty-five cents an hour. Leavitt worked seven hours a day, six days a week. She developed a good reputation even among the talented staff at the observatory—Margaret Harwood would later remark that Leavitt had "the best mind at the Observatory"—and was charged by the director, Edward Pickering, with studying the plates sent from Harvard's Southern Hemisphere observatory in Peru to look for interesting stars. In looking over the plates, Leavitt noticed an intriguing pattern in a class of stars known as Cepheid variables. Her observations would end up radically transforming our understanding of the universe.

As the name suggests, Cepheid variables are stars whose intrinsic brightness changes in time; the first such star studied in detail was located in the constellation Cepheus, which supplies the other half of the name. The intensity variation is now known to be driven by an oscillation in the size of the star—the outer layers of gas pulse in and out in a regular way,

---

* The same thing had happened to Annie Cannon, coincidentally, and the two worked together for a time.

heating up and expanding, then cooling and contracting. As the star expands, it brightens, and as it contracts, it dims.

Leavitt took an interest in these variable stars and identified over twenty-four hundred of them in the course of her observations.* While doing this work, she noticed a pattern in the oscillation: The brighter Cepheid variables in her images seemed to take longer to go from bright to dim and back again. On her own initiative, she set out to test whether this apparent pattern was real.

Of course, there are many confounding factors when it comes to establishing the actual brightness of stars, chief among them being the effect of distance. When we look at the light from the sky, we see a two-dimensional projection of a three-dimensional universe, with no easy way to establish the distance along our line of sight.† A dim object on the sky might be dim because it intrinsically has a low luminosity, or it might in fact be a very bright star that is simply very far away from us.

To get around this problem, Leavitt concentrated on some twenty-five Cepheid variables in the Small Magellanic Cloud. This is now known to be a small satellite galaxy orbiting our Milky Way; at the time, all that was known was that it was a large group of very distant stars, sufficiently distant that the

---

* Even today, about one in ten of the well-studied variable stars known to astronomers were first studied by Henrietta Leavitt.

† We estimate distances to everyday objects by comparing the slightly different images from our left and right eyes; the apparent position of an object will change from one eye to the other, with the effect being most pronounced for nearby objects. You can demonstrate this by holding a pen out at arm's length, and looking at it with first one eye closed, then the other. The distances to other stars are so great, however, that even comparing the apparent position of stars seen from opposite ends of the Earth's orbit produces an exceedingly small change. Although astronomers looked for this *stellar parallax* starting around the time of Copernicus, the first successful measurement wasn't made until the 1800s.

depth of the cloud is insignificant compared with the distance to the cloud. For all practical purposes, these twenty-five stars were all the same distance from Earth, and thus their relative brightness could safely be assumed to reflect real differences in the intrinsic luminosity of the stars and not mere distance.

Within this set of comparable stars, Leavitt's painstaking observations showed a clear relationship between the period of the variation and the luminosity of the star. The brighter the star, the longer the period, in keeping with a clear and simple mathematical formula. This might seem a mere curiosity, but in fact it was a revolutionary discovery, because it gives a way to establish the intrinsic luminosity of a Cepheid variable from simply measuring the period of variation. And if you know both the intrinsic luminosity and how bright the star *appears* to be, you can easily determine the distance to Earth.

Leavitt's results were published (under Pickering's name, but the work was widely known to be hers) in 1912. In 1913, Ejnar Hertzsprung measured the distance to a handful of Cepheid variables in the Milky Way. Putting these two results together enables Cepheid variables to serve as "standard candles," allowing accurate distance measurements over the scale of galaxies. Harlow Shapley combined these results to establish the distance to some globular clusters, setting the scale of the Milky Way. Then, in 1924, Edwin Hubble used Leavitt's relationship to radically transform our understanding of the scale of the universe. Hubble spotted several Cepheid variables in what was then known as the Andromeda spiral nebula, whose brightness allowed him to measure the distance to Andromeda at more than a million light-years. The distance lay well outside Shapley's measurement of the bounds of the Milky Way and established conclusively that Andromeda was not, in fact, a nebulous collection of gas within our own galaxy, but a separate galaxy in its own right. Hubble's findings on Andromeda ruled out the theory, favored by Shapley, that our galaxy represented the entire universe. The Milky Way now became merely one of a vast number of galaxies separated by mind-boggling distances.

Hubble wasn't done using Leavitt's discovery to transform astronomy, though. In 1929, he announced another incredible discovery: Using Cepheid variables (among other methods) he showed a clear relationship between the distance to a galaxy and the rate at which it appears to be moving away from us. "Hubble's law," clear evidence that the universe as a whole is expanding, led directly to the modern Big Bang cosmology, the idea that the universe began as a single hot, dense point some 13.7 billion years ago and has been expanding and cooling ever since. Numerous other lines of evidence have since been found supporting the Big Bang model, but it begins with Hubble's relationship between distance and speed.

Our whole understanding of the scale and history of the universe, then, can be traced directly to the pattern-spotting abilities of Henrietta Swan Leavitt, poring over photographic plates for twenty-five cents an hour. Her ability to pick out an interesting pattern from a vast array of visual data was the key that enabled all the later discoveries.*

THE DEATH AND AFTERLIFE OF STARS

Leavitt's work, like all astronomy up until the twentieth century, used the visible light coming from stars. In 1931, however,

---

* In recognition of Leavitt's great contribution to astronomy, the Swedish mathematician Gösta Mittag-Leffler wanted to nominate her for the 1925 Nobel Prize; alas, she had died of cancer in 1921, and the Nobel is not awarded posthumously.

a radio engineer named Karl Jansky introduced an entirely new tool for studying the cosmos. Charged with identifying potential sources of interference with radio communications, Jansky had built a large rotatable antenna that he could point in all directions. In the process, he discovered an intense source of radio waves (which are just light with an extremely long wavelength) coming from the direction of the constellation Sagittarius and associated with the center of the Milky Way galaxy.

Jansky's discovery marks the birth of radio astronomy, though it took a while to become a major field. The intense research on radar during World War II greatly improved the technology for detecting and analyzing radio waves, though, and in the postwar decades, astronomers began to conduct serious studies of the radio waves coming from astronomical sources.

Working at radio wavelengths offers both advantages and disadvantages for astronomy. On the one hand, studying invisible radio waves requires more complicated detectors than does studying visible light. On the other hand, the long wavelength of radio waves means that they're relatively forgiving in terms of the construction of detectors. A telescope mirror for visible light, with a wavelength of a few hundred nanometers, needs to be polished smooth at a level of a few tens of nanometers to form a high-quality image. A radio telescope built to detect light with a wavelength of tens of centimeters, however, can tolerate much greater imperfections.

This greater tolerance for deviation makes it easier to construct truly enormous radio telescopes. One of the earliest of these was the Interplanetary Scintillation Array assembled by Antony Hewish and his students in the mid-1960s. It consisted of an array of wires strung between posts sledgehammered into the ground, covering a 4½-acre field a few kilometers from Cambridge, England.*

One of the students pounding posts into the ground was Jocelyn Bell (later Bell Burnell), who was later given the task of monitoring the output of the telescope. The initial survey was essentially passive, simply recording all the signals that came in as the instrument gradually swept across the entire sky as the Earth's rotation carried the telescope along. The output was hooked up to pens tracing squiggly lines on a chart recorder, running through around a hundred feet of paper a day during normal operation. Bell's job was to look through the miles of paper as it came in, looking for both man-made interference and interesting astronomical sources. One day in the summer of 1967, Bell noticed a bit of "scruff" on the chart—a region where the incoming signal was varying very rapidly. The unusual signal lasted a few minutes and covered maybe a quarter inch of paper. This didn't match any of the sources Bell's group was looking for, but she recalled seeing something similar earlier. Searching back through the hundreds of feet of paper, she found the same "scruff" turning up at regular intervals, which indicated that it was associated with a particular position in the sky.

Closer study—which involved Bell's coming in at the appropriate time and increasing the speed of the chart recorder

---

* For comparison, the largest visible-light telescopes in the world have a diameter of about ten meters, corresponding to an area of 0.02 acres. The largest single radio telescope in the world is the Arecibo Observatory, with an area of 118 acres. Radio astronomers can also combine the signals from multiple telescopes electronically, with the combined instrument functioning like a single huge telescope; the largest such projects use telescopes on different continents to make a telescope effectively the size of the entire Earth.

paper for a few minutes in the hope of catching the signal again—revealed that the unusual signal was an extremely regular series of pulses, one every 1.339 seconds. The regularity of the signal at first seemed artificial, but careful investigation ruled out any earthbound source. For a brief time, the researchers referred to it as an "LGM" signal, for "Little Green Men," but when a second such source, with a different period, turned up on the opposite side of the sky, they realized it must have a natural origin.

The strength and regularity of the signal indicated that whatever the source was, it had to contain an enormous amount of mass and energy—the amount of energy being beamed away would rapidly deplete anything smaller. The rapidity of the repetition suggested that the source had to be very compact—the most likely natural cause of a repeating signal was a rotating object, and such rapid rotation would be implausible for something the size of a star. In 1968, it was suggested that pulsars were, in fact, rapidly rotating neutron stars, remnants of supernova explosions packing a mass greater than that of the Sun into a sphere only tens of kilometers across. Further support for this theory was provided by the detection of a pulsar in the Crab Nebula, the remnant of a supernova that was visible on Earth in 1054 CE.

Neutron stars get their name because they pack so much mass into such a small space that the electrons and protons making up ordinary matter merge, and almost the entire mass of the star consists of neutrons. Only the quantum-mechanical prohibition on identical particles occupying the same state prevents the neutron star from collapsing into a black hole, an infinitesimal point where the ordinary laws of physics break down altogether. The pulsing that caught Bell's attention is due to rapidly moving particles near the star's magnetic poles, which generate an intense beam of radiation.* As the neutron star rotates, the beam sweeps across the sky like a lighthouse beam and is detected as a short pulse of light during the brief period when it's pointed directly at Earth.

The possibility of black holes had been recognized as a consequence of Einstein's general theory of relativity within a few months of its publication in 1915, and neutron stars were proposed within a couple of years of the discovery of the neutron. Whether such extreme objects could actually exist or were merely a mathematical curiosity remained a topic of intense debate among astronomers for many years. Bell's discovery of pulsars proved that these exotic stars were, in fact, real objects, and the existence of neutron stars and black holes is now generally accepted.

Since 1967, around two thousand pulsars have been detected and studied. Observations of binary systems in which a smaller, visible star orbits a neutron star confirm the large mass and small size of the neutron stars. Small changes in the rotation rate of pulsars provide information about the internal structure and behavior of the neutron stars themselves—"quakes" caused by internal rearrangements lead to sudden jumps in the rotation rate, while a steady slow decrease in the rate provides a measure of the rate at which energy flows out into space. A few binary systems consisting of two neutron stars have been observed and provide the most convincing evidence we have of gravitational waves, another of the exotic predictions of general relativity. Randall Hulse and Joseph Taylor shared the 1993 Nobel Prize in Physics for discovering the first such system.

---

* Pulsars have been detected in many wavelength ranges, from radio to gamma rays. The Crab Nebula pulsar visibly flickers about thirty times per second as it rotates.

In the forty-six years since the discovery of pulsars, these objects have provided a vast trove of information about the death and afterlife of stars. They are our best source of information about the behavior of matter under extreme conditions, at the limits of our current theories. And all of this can be traced back to a grad student who happened to spot an unusual bit of "scruff" in miles of chart recorder traces.[†]

GALAXY ZOO

The stories of Henrietta Swan Leavitt and Jocelyn Bell Burnell illustrate the important role of pattern recognition by trained astronomers. In 2007, a different kind of project was launched, one that takes advantage of Internet technology to leverage the pattern-recognition abilities of untrained volunteers on a huge scale.

As in Leavitt's day, the development of new detector technology has inspired surveys of wide swaths of the sky, not with photographic plates, but with charge-coupled device (CCD) cameras producing digital images that are more easily manipulated. The Sloan Digital Sky Survey collects detailed images of areas covering more than a quarter of the sky in several wavelengths of light, making it one of the most ambitious astronomical surveys in history. Over two hundred million

---

[†] Bell Burnell's boss, Antony Hewish, shared the 1974 Nobel Prize in physics for the discovery, but she was not included. This oversight is widely regarded as one of the worst slights in Nobel history and an example of pervasive sexism. Bell Burnell herself is a little more sanguine, attributing it to a different long-standing problem, that of senior researchers' getting the credit for work done by their students. In a 2009 interview, she said, "I have discovered that one does very well out of not getting a Nobel Prize, especially when carried, as I have been, on a wave of sympathy and a wave of feminism" (Douglas Colligan, "The Discover Interview: Jocelyn Bell Burnell," *Discover*, November 2009).

celestial objects have been identified in these images to date, including a huge number of galaxies.

Since the study of galaxies began in the early 1900s, astronomers have classified them into several types within two general groupings. *Spiral galaxies* have distinct bright and dark "arms"; the Andromeda galaxy is the nearest and best known of these and serves as the default mental image most people have of a galaxy. *Elliptical galaxies* appear to be merely large collections of stars, without obvious structure. These classifications are an important part of research on galaxy evolution.

Kevin Schawinski, a grad student working with the Sloan data, needed large numbers of galaxy classifications for his Ph.D. research. Classifying galaxies is a task in which humans vastly outperform computers, but is an extremely time-consuming process for a single astronomer. Schawinski personally classified some fifty thousand images, but it required a week of intense effort and long hours.

To speed up the process, Schawinski and a postdoc on the project, Chris Lintott, realized that the classification they needed doesn't require extensive training; it just needs the innate pattern-matching ability of a human brain. So they turned to the Internet to recruit volunteer classifiers, launching the Galaxy Zoo project.

Visitors to the initial Galaxy Zoo web page were presented with an image of a galaxy and asked to answer two simple questions: Was the galaxy in the picture spiral or elliptical, and if it was a spiral, did the arms spiral clockwise or counterclockwise? Each galaxy was shown to several human volunteers, with the different ratings collected and compared to provide both a classification of the object and some estimate of the confidence in that classification (based on how well the classifications agreed with one another).

Schawinski and Lintott initially expected it might take two years to go through their data set of roughly one million galaxies. In the first year of operation, though, they received more than fifty million classifications from some 150,000 human volunteers. A follow-on project, Galaxy Zoo 2, asked more-detailed questions about some quarter million of the brightest galaxy images and received more than sixty million responses in just fourteen months. Not only were people willing to look at these images on a lark, but large numbers of people found it fun and spent hours on the site. Looked at just in terms of the original goal, the project was a smashing success.

But classification alone isn't the whole story. As happened with Leavitt and Bell, just having humans look over this vast archive of images led to unexpected discoveries. Even without a background in astronomy, humans excel at spotting odd patterns that would go unnoticed by computers.

About a month after the launch of Galaxy Zoo, on August 13, 2007, a Dutch schoolteacher named Hanny van Arkel noticed an odd greenish blob beneath a galaxy she had been asked to classify. In addition to providing a simple platform for showing images and getting classifications from volunteers, the Galaxy Zoo project included a forum in which volunteers could discuss the images they were working on, so van Arkel posted a message asking if anybody knew what the blob was.

Nobody recognized it, including the professional astronomers monitoring the project. The puzzle of Hanny's voorwerp (after a Dutch word for "object") prompted several astronomers to begin follow-up studies, eventually including observations from the Hubble telescope and the Suzaku and Newton x-ray satellite telescopes and various ground-based optical and radio telescopes. The voorwerp turns out to be a cloud of gas tens of thousands of light-years across, which has somehow been heated to high enough temperatures to emit visible light. The exact mechanism of the heating is still under debate, but probably involves a gigantic black hole at the center of the galaxy that was the original target. Hanny van Arkel has coauthored a half-dozen papers about her discovery with scientists from the Galaxy Zoo team.

In addition to the voorwerp, the Galaxy Zoo project also discovered a second new class of objects, dubbed "Green Peas" because they show up in the images as compact green balls. These are particularly interesting from the standpoint of citizen science, because unlike what happened with the voorwerp, the professional astronomers on the project did not become involved very quickly. Instead, the identification and initial analysis of the Green Peas was carried out by enthusiastic amateurs, some of whom became quite adept at interpreting the available data on the Green Peas. Study of these interesting objects has produced another significant collection of Galaxy Zoo papers.

All told, the Galaxy Zoo and Galaxy Zoo 2 projects had led to more than thirty published scientific articles as of early 2013. These include articles on the originally intended purpose—looking at the distribution of spiral and elliptical galaxies—as well as wholly new discoveries like the voorwerp and the Green Peas. Leveraging the pattern-spotting abilities of thousands of citizen scientists has proven to be a very productive tool for science, and the Zoo has expanded to include a wide range of projects.

THE ZOONIVERSE

Most of the chapters of this book conclude with examples of ways you can employ lessons from science in your everyday life and activities, but this chapter is slightly different. Rather than offering a way to use scientific thinking to better your life, I'd like to offer an invitation to use your everyday thinking to help science.

In 2009, the Galaxy Zoo team expanded its focus to provide a platform for a wider range of projects, using the same citizen-science template: Volunteers go to the website (www. zooniverse.org), choose an interesting-sounding project, and are presented with data of one sort or another and asked to spot a pattern. By combining the efforts of thousands of volunteers—nearly eight hundred thousand people have registered with the Zooniverse—Galaxy Zoo helps researchers find crucial signals in the sorts of data where human brains are superior to computers. As of November 2013, there were nineteen active projects on the Zooniverse site, covering a range of different sciences, plus a handful of "retired" projects.

Given the project's origins in astronomy, it's not surprising that the bulk of the projects involve astronomical data. Three of these were direct outgrowths of the original Galaxy Zoo project: a continuation of the galaxy-classification project, now working with images from the Hubble telescope archives; a search for supernovae in galaxy images; and a study of merging galaxies, now completed, which asked volunteers to compare the results of simulated galaxy collisions to images of actual colliding galaxies. This last task, the so-called Merger Wars project, is another example of studies where the snap judgments of humans can make progress faster than computer algorithms. The images are very complicated, making a detailed comparison on a computer extremely difficult, but a human looking at the whole image can judge the similarity of the images with relative ease.

Other astronomy-themed projects ask volunteers to identify circular patterns in photographs: "bubbles" in infrared images of gas and dust that indicate the presence of newly forming stars in our galaxy (the Milky Way project) or craters on the Moon (Moon Zoo). Others track weather elsewhere in the solar system, either through spotting windblown dust in photographs of Mars (Planet Four) or solar storms in short video clips of the outer atmosphere of the Sun (Solar Stormwatch).

The Planet Hunters project does not use photographs, but rather uses data showing variation over time, like the charts studied by Jocelyn Bell. Volunteers look for small decreases in the brightness of one of the 150,000 stars monitored by the Kepler satellite. The dimming would be caused by a planet passing between the star and the Earth. These dips can be rather subtle, making it difficult for computers to sort them out from natural fluctuations in a given star's brightness, but humans have proven to be fairly adept at this. Over its first three years of operation, Planet Hunters logged more than twenty million volunteer classifications, including the identification of more than thirty likely planets.

The range of available projects extended beyond astronomy. There are zoological investigations, such as a project that asks volunteers to identify animals in photos of the sea floor (Seafloor Explorer). Other projects have volunteers identify images of tiny ocean life (Plankton Portal) or images from "camera traps" in Africa (Snapshot Serengeti). One microbiological project asks participants to look for cancer cells (Cell Slider), and another for egg-laying worms (Worm Watch Labs). Still other projects exploit a different highly developed visual skill, the ability to read bad handwriting: Volunteers decipher weather records from merchant ship logs (Old Weather), museum records (Notes from Nature), or Greek letters in ancient papyrus fragments (Ancient Lives). A different sense altogether is the shared skill for some other projects, which ask volunteers to identify patterns in audio recordings of animal calls (Whale FM, Bat Detective).

There are Zooniverse projects to suit all kinds of skills and interests, and all of these projects rely on tasks that are fairly trivial for human volunteers. In most cases, it takes no more than a few seconds to evaluate a given candidate. Some of the tasks are also weirdly addictive—clicking through dozens of light curves in Planet Hunters has proven to be a fantastic way for me to procrastinate on any number of writing projects. None of the Zooniverse projects require any particular background knowledge—you don't need to be able to read Greek to match up characters in Ancient Lives, or know much about astrophysics to categorize galaxies—though participating in the projects will often lead to your learning something about the subject almost by accident. I know a good deal more about the operation of stars than I did prior to researching these projects, just from reading the discussion pages about some of the odder light curves in Planet Hunters.

The Zooniverse is one of the largest collections of citizen-science projects, but there are many other options, both single labs and large collections. These range from almost completely passive projects like SETI@Home and Einstein@Home, which use the idle time of participants' computers to run automated routines sifting through large data sets, to projects requiring a high level of engagement, like the array of bird-watching activities operated by the Cornell Lab of Ornithology. There are even more commercial operations, like Dognition, which offers a "personality test" for your pet as a way of amassing a large set of data for researchers at Duke University studying the mental abilities of dogs. The SciStarter.com website offers a large index of projects from many organizations.

Several of these projects also incorporate game-like elements, offering rankings and achievements to be unlocked. These games involve a fairly minimal investment of effort, comparable to what many people put into FarmVille, Candy

Crush, and other social-media games, though vastly less irritating to your friends and relations on Facebook. Using these scientific interfaces and the unparalleled pattern-matching ability of your own brain, you can help advance the progress of science during your spare moments at any computer with an Internet connection.

STEP TWO

THINKING

The sciences do not try to explain, they hardly even try to interpret, they mainly make models.

—John von Neumann, "Method in the Physical Sciences,"
in *The Unity of Knowledge*, edited by L. Leary

Once a set of observations or experimental results has been collected, the next step in the process of science is to think of an explanation for those observations or results. The products of this thinking step are generally referred to as *models*.

A good scientific model provides a story explaining not just what happened in a given situation, but *why* it happened. Successful models allow scientists not only to explain past events, but also to predict the results of future observations.

Scientists try to base their models on the most fundamental and universal principles possible. A model of the universe that relies on the particular properties of specific objects or the direct intervention of capricious supernatural beings has very little predictive value. While you could make a model of the solar system that has Earth orbiting the Sun more quickly than Mars because invisible gravity elves like Earth better, the model doesn't let you say anything about the orbits of Jupiter or Venus. Isaac Newton's theory of gravity, on the other hand, explains the orbits in terms of a universal attraction between planets and the Sun, which correctly predicts the orbital speeds of not just Earth and Mars, but also all of the other planets and the asteroids and comets in the solar system.

Scientific models are built up from observations, and the success of a model is ultimately determined by future observations and experiments. The model-building process itself can be fascinating and involves a remarkable amount of inspiration and creativity. In this section, we'll look at some examples of everyday activities that involve putting together models from observations, and we'll present some historical examples of scientific discoveries that draw on the same model-making process.

Chapter 5

SETTING THE (PERIODIC) TABLE

DICK: I guess it looks as if you're reorganizing your records.
What is this though? Chronological?
ROB: No . . .
DICK: Not alphabetical . . .
ROB: Nope.
DICK: What?
ROB: Autobiographical. [. . .]
DICK: That sounds very . . .
ROB: Comforting.

—*High Fidelity*
(2000 film starring John Cusack)

I met Kate via the Internet, specifically a couple of Usenet groups discussing science fiction and fantasy books.* These groups were frequented by lots of people who owned lots of books, and there were fairly regular discussions about

---

* Usenet was a text-based discussion forum whose popularity peaked in the mid- to late 1990s, when I was a graduate student in Maryland. Dating a woman I met via the Internet led to no small amount of teasing from my less geeky friends; making matters worse, we started dating in 1998, when she was a congressional intern and the Lewinsky scandal was in full swing.

the various schemes people used to shelve their collections. Some people separated their collections by subject and literary genre; others lumped together different genres, even fiction and nonfiction. Alphabetical by author was probably the most common method of organization within categories, but some people preferred other arrangements. A few folks even employed schemes that were essentially random—chronological by date of purchase, or even deliberately scattering books around the shelves—on the grounds that having to scan through all the shelves to find a particular book to read encouraged serendipity.

The most complicated scheme I recall was employed by another friend from Usenet. He sorted books first by binding (hardcover versus paperback), then by trim size (larger hardcovers on a separate shelf from smaller ones), then by publisher, and only then by author. The goal of all this was to ensure that each shelf contained a complete row of books with the publishers' logos on the spine neatly aligned. Mike found this very aesthetically satisfying, but the rest of us thought it would just make it impossible to find anything.

This kind of behavior is fairly typical of collectors. Once you've amassed a large number of collectible items, the next step is figuring out how best to sort them so as to provide an aesthetically pleasing display, or to make it easy to find things quickly, or for whatever other purpose suits your interests. Many collectors take pleasure in periodically re-sorting their collections using new classification schemes. Sometimes this is for comfort in bad times—like John Cusack's record collector, Rob, in the above-quoted scene from *High Fidelity* (or the Nick Hornby novel that it's based on), who responds to a bad breakup by sorting his albums "autobiographically"—or just to have an excuse to go through the collection and look at everything again.

This sorting impulse can seem obsessive from the outside, but it can also serve as a starting point for science. Once you've amassed a large number of observations about the world, the next step is to figure out how they all fit together. Deciding how to arrange the data often helps make underlying patterns clear, ultimately leading to a deeper understanding. The most famous example of sorting as a path to discovery is also one of the most iconic images of science, the periodic table of the elements (Figure 5.1).

The periodic table is one of the most reliable visual markers of a science classroom, a vaguely rectangular array of boxes, with two towers rising up at either end, and a little island of two rows floating off by itself at the bottom. Each box represents one of the 118 known chemical elements, which combine to form all the material objects we interact with in the everyday world. The elements are grouped together in rows and columns, and some essential chemical properties can be instantly determined from an element's position on the periodic table, in particular the number of other atoms it will form chemical bonds with.

The general idea of the periodic table has gained a foothold of sorts in popular culture, and you can find "periodic tables" of all kinds of different items on the Internet—periodic tables of meats, condiments, beer, cars, authors of various literary genres. Google will helpfully turn up a "periodic table" of just about any collective noun you might think of, in addition to innumerable variations on the actual periodic table of the elements, such as the Comic Book Periodic Table, maintained by two chemists at the University of Kentucky, which collects comic book panels mentioning many of the elements.

You can also buy any number of periodic-table-themed products—T-shirts, coffee cups, shower curtains. There are

Figure 5.1. The periodic table.

Source: "Periodic-table," Wikimedia Commons, last updated September 3, 2011, http://commons.wikimedia.org/wiki/File:Periodic-table.jpg.

hydrogen 1 H 1.0079																	helium 2 He 4.0026
lithium 3 Li 6.941	beryllium 4 Be 9.0122											boron 5 B 10.811	carbon 6 C 12.011	nitrogen 7 N 14.007	oxygen 8 O 15.999	fluorine 9 F 18.998	neon 10 Ne 20.180
sodium 11 Na 22.990	magnesium 12 Mg 24.305											aluminium 13 Al 26.982	silicon 14 Si 28.086	phosphorus 15 P 30.974	sulfur 16 S 32.065	chlorine 17 Cl 35.453	argon 18 Ar 39.948
potassium 19 K 39.098	calcium 20 Ca 40.078	scandium 21 Sc 44.956	titanium 22 Ti 47.867	vanadium 23 V 50.942	chromium 24 Cr 51.996	manganese 25 Mn 54.938	iron 26 Fe 55.845	cobalt 27 Co 58.933	nickel 28 Ni 58.693	copper 29 Cu 63.546	zinc 30 Zn 65.39	gallium 31 Ga 69.723	germanium 32 Ge 72.61	arsenic 33 As 74.922	selenium 34 Se 78.96	bromine 35 Br 79.904	krypton 36 Kr 83.80
rubidium 37 Rb 85.468	strontium 38 Sr 87.62	yttrium 39 Y 88.906	zirconium 40 Zr 91.224	niobium 41 Nb 92.906	molybdenum 42 Mo 95.94	technetium 43 Tc [98]	ruthenium 44 Ru 101.07	rhodium 45 Rh 102.91	palladium 46 Pd 106.42	silver 47 Ag 107.87	cadmium 48 Cd 112.41	indium 49 In 114.82	tin 50 Sn 118.71	antimony 51 Sb 121.76	tellurium 52 Te 127.60	iodine 53 I 126.90	xenon 54 Xe 131.29
caesium 55 Cs 132.91	barium 56 Ba 137.33	57-70 *	lutetium 71 Lu 174.97	hafnium 72 Hf 178.49	tantalum 73 Ta 180.95	tungsten 74 W 183.84	rhenium 75 Re 186.21	osmium 76 Os 190.23	iridium 77 Ir 192.22	platinum 78 Pt 195.08	gold 79 Au 196.97	mercury 80 Hg 200.59	thallium 81 Tl 204.38	lead 82 Pb 207.2	bismuth 83 Bi 208.98	polonium 84 Po [209]	astatine 85 At [210]	radon 86 Rn [222]
francium 87 Fr [223]	radium 88 Ra [226]	89-102 **	lawrencium 103 Lr [262]	rutherfordium 104 Rf [261]	dubnium 105 Db [262]	seaborgium 106 Sg [266]	bohrium 107 Bh [264]	hassium 108 Hs [269]	meitnerium 109 Mt [268]	110 Uun [271]	111 Uuu [272]	112 Uub [277]		ununquadium 114 Uuq [289]				

*Lanthanide series

lanthanum 57 La 138.91	cerium 58 Ce 140.12	praseodymium 59 Pr 140.91	neodymium 60 Nd 144.24	promethium 61 Pm [145]	samarium 62 Sm 150.36	europium 63 Eu 151.96	gadolinium 64 Gd 157.25	terbium 65 Tb 158.93	dysprosium 66 Dy 162.50	holmium 67 Ho 164.93	erbium 68 Er 167.26	thulium 69 Tm 168.93	ytterbium 70 Yb 173.04

** Actinide series

actinium 89 Ac [227]	thorium 90 Th 232.04	protactinium 91 Pa 231.04	uranium 92 U 238.03	neptunium 93 Np [237]	plutonium 94 Pu [244]	americium 95 Am [243]	curium 96 Cm [247]	berkelium 97 Bk [247]	californium 98 Cf [251]	einsteinium 99 Es [252]	fermium 100 Fm [257]	mendelevium 101 Md [258]	nobelium 102 No [259]

multiple books on the periodic table, both in popular science writing and in literature, where Italian chemist and Holocaust survivor Primo Levi famously used it as the organizing principle for an autobiographical story collection.* Several companies sell periodic table building blocks, if you want to get your toddlers started early on organizing elements into rows and columns. And if you have a spare eighty-six hundred dollars lying around, you can even buy a periodic table coffee table, complete with samples of all ninety-two naturally occurring elements embedded in acrylic.

These days, the organizing principle underlying the periodic table is well understood in terms of the internal structure of atoms. Amazingly, though, the periodic table long predates modern atomic theory; in fact, when the table was first developed in the late 1860s, there was still considerable debate about whether atoms were real entities at all, or merely a calculational convenience. The periodic table as we know it today comes from the sorting impulse of a Russian chemist who found himself with a large collection of information and desperately needed a way to organize it all.

A TEXTBOOK EXAMPLE

When Dmitrii Ivanovich Mendeleev arrived in St. Petersburg in 1861 after two years of study in Heidelberg, Germany, he found himself in a situation familiar to many modern college

---

* Among the popular-science choices, my local bookstore offers *The Disappearing Spoon* by Sam Kean, *Periodic Tales* by Hugh Aldersly-Williams, several coffee-table books containing glossy photos of all the elements, and a charming illustrated version from Japanese artist Bunpei Yorifuji representing each element and its properties by a little cartoon figure (heavy elements are fat, artificially made elements are robots, etc.).

graduates: on his own in a big city with no job and with loans to repay.* Despite good academic credentials, he couldn't immediately secure a teaching position, because he had arrived in the middle of the year. He badly needed a way to make some money to support himself and pay off the thousand rubles he had borrowed for his studies in Germany.

Mendeleev's solution to this problem was rather unusual: He decided to write a textbook on organic chemistry. Not only did he find a publisher to pay him for the book, which he finished in a matter of months, but he also won the Demidov Prize of the Petersburg Academy of Sciences in early 1862. The prize committee consisted of two of his academic mentors, so in a sense the fix was in, but the book was still a notable achievement. As the first organic chemistry textbook in Russian—previous generations of students had to use translated German texts—it was an immediate hit with professors and students and was much admired.

Having made an academic reputation with his textbook, Mendeleev secured a teaching job at the Technological Institute in Petersburg in 1863 and then took a position at St. Petersburg University in 1867. On making the move to the university, he took up teaching its general inorganic chemistry course and decided to write another textbook to accompany his lectures. He signed a contract with a publisher for a two-volume text, *The Principles of Chemistry*, and began work in 1868.

Inorganic chemistry is a catchall term covering essentially all of the elements except carbon, hydrogen, oxygen, and

---

* Being Russian, Mendeleev's name properly belongs in the Cyrillic alphabet and has been transliterated into the Latin alphabet in a dizzying variety of ways. I'm going with the version used by Michael D. Gordin, *A Well-Ordered Thing: Dmitrii Mendeleev and the Shadow of the Periodic Table* (New York: Basic Books, 2004).

nitrogen. In Mendeleev's time, there were sixty-three known elements (with chemists of the day working feverishly to discover more), and describing all their properties was no trivial task. Mendeleev further complicated his task by structuring the first volume around the practices of chemistry, rather than the elements themselves. His introductory chapters discuss extremely general principles, natural history, and simple experiments, then the chemistry of water and its properties. This approach was very effective pedagogically, but twenty chapters in, he had discussed the properties of only four elements: hydrogen, oxygen, carbon, and nitrogen—those important for organic chemistry. The twentieth chapter begins with a discussion of table salt (sodium chloride) and then discusses the properties of chlorine and the other halogens (fluorine, iodine, bromine), a group of elements that share many properties with chlorine. Halogens form colored vapors, react very strongly, and combine with hydrogen to make powerful acids. That brings to eight the total number of elements discussed in the first volume (sodium was left for the second volume).

Mendeleev thus faced a classic problem of nonfiction writing: He had a mountain of information to present (fifty-five of the sixty-three known elements), but had limited space in which to do it. To complete his textbook in the contractually mandated two volumes, he needed a more efficient way to present his material. Accordingly, he began casting around for a way to group the elements together, so as to dispose of several per chapter.

The halogens, discussed at the end of the first volume, provided a clue, as did the alkali metals, a group of elements (lithium, potassium, rubidium, and cesium) with properties very similar to those of sodium: They're all highly reactive metals, which famously burst into flame when they come in contact with water.* The alkaline earths (beryllium, magnesium, calcium, strontium, barium) are another group with some obvious chemical similarities.†

The distinctions between many of the other elements are far more subtle—if an insane chemist ever puts a gun to your head and asks you to give one property of an element you've never heard of, "It's a greyish metal" correctly describes a huge swath of the periodic table. Mendeleev nevertheless began writing down all the known properties of all the elements and soon found a very general pattern: Certain characteristics tend to recur at regular intervals when you arrange the elements in order of their atomic weight.‡ The pattern is easiest to see for the halogens and alkali metals, two groups of chemically similar elements.

The halogens and their atomic weights are as follows:

Symbol:	F	Cl	Br	I
Name:	Fluorine	Chlorine	Bromine	Iodine
Weight:	19	35.5	80	127

---

* Laser cooling, the subfield of atomic physics on which I did my graduate research, got its start with studies of alkali metals, and rubidium is probably the most widely used element for laser-cooling experiments. As a result, most physicists in the field have stories about nearly setting their labs or themselves on fire, thanks to the reactions of alkali metals. These make a reliably entertaining topic of conversation in bars at conferences.

† The modern periodic table includes radium among the alkaline earths, but its discovery was still thirty years off when Mendeleev was making his table.

‡ One popular story about Mendeleev's sorting has him writing all the elements on index cards, then shuffling them around into various patterns. Charming as the idea is, there's no contemporary evidence that he did this—the first mentions of sorting cards come decades later.

And the alkali metals:

Symbol:	Li	Na	K	Rb	Cs
Name:	Lithium	Sodium	Potassium	Rubidium	Cesium
Weight:	7	23	39	85	133

The first members of each of these groups are separated by 16 or 17 units in atomic weight, with later members separated by 46 to 48 units. There's also a clear relationship between the groups: Sodium is 4 units heavier than fluorine, potassium is 3.5 units heavier than chlorine, rubidium is 5 units heavier than bromine, and cesium 5 units heavier than iodine. The pattern can also be extended to the alkaline earths, where each element is between 1 and 5 units heavier than the corresponding alkali metal.

Mendeleev found similar regularities in numerous other groups of elements and gradually became convinced that these patterns were not just pedagogically convenient, but reflected a deep principle, which he termed the "periodic law." In February 1869, he sent the first draft of his periodic table to his publisher and, the next month, sent a paper to the Russian Chemical Society. He continued to refine his system, publishing a paper on it in Russian in 1870, and then in German in 1871, which finally brought his system to the attention of the rest of European science.§ The table in the 1871 paper incorporated all the elements known at the time and, crucially, a few that were yet to be discovered, and secured his place as the father of the periodic table we know and love today.

---

§ The 1870 paper was actually presented at the Russian Chemical Society meeting by a friend on his behalf, as Mendeleev was out of town inspecting artisanal cheese-making operations for the Russian government.

PERIODICITY AND PREDICTION

As was the case with Darwin and evolution (see Chapter 1), Mendeleev was not the only chemist thinking about schemes for organizing the elements in the late 1860s. The English chemists William Odling and John Newlands had suggested periodic principles for organizing the chemical elements in 1864, with Odling making a table that looked very much like Mendeleev's first attempt in 1869. The French chemist Alexandre-Emile Béguyer de Chancourtois had an even earlier scheme, in 1862, placing the various elements on a line spiraling around a cylinder. The spiral system clearly involves periodic behavior, but de Chancourtois's paper was difficult to follow and not widely known. The German chemist Lothar Meyer produced a table nearly identical to Mendeleev's in 1870, including a graph displaying the atomic volume as a function of atomic weight, a dramatic visual demonstration of periodic characteristics. Meyer and Mendeleev subsequently had a long-running dispute about who was really the first to develop a periodic table.

As with evolution, it seems that periodic ideas were in the air in the 1860s. Mendeleev, however, while not exactly a household name, is vastly better known than any of his competitors and generally gets the credit for being the father of the periodic table. So why is he remembered, while the others are mostly forgotten?

Mendeleev gets the most credit because in terms of the scheme I'm using to classify stories in this book, he straddles the line between looking and thinking. His table was based on observations of patterns in the known elements, but having noticed those patterns, he developed the idea of a periodic law and used that theoretical model to make predictions, something not matched by his contemporaries. He was sufficiently convinced of the periodic law that he even used it to "correct" the properties of some elements that didn't fit perfectly within his scheme. Sometimes this was successful, as when he asserted that the then accepted value for the mass of beryllium, about 14 units, had to be wrong because it wouldn't fit properly within his system. Mendeleev argued that its mass should instead be around 9 units, which later measurements showed to be correct.

In other cases, his belief in the accuracy of his system led Mendeleev astray. He argued that the atomic mass of tellurium, which is greater than that of iodine, must have been incorrectly measured, as tellurium needed to come before iodine in the periodic system, in light of the two elements' chemical properties. He was right about the chemical properties—iodine belongs with the halogens, after tellurium in the modern ordering—but wrong about the mass. Tellurium is, indeed, more massive than iodine, but this does not invalidate the periodic law; the observation only indicates that Mendeleev wasn't using exactly the right property to arrange his list of elements (more about this later).

Mendeleev's great triumph, though, came with the "eka-elements." When he went through the known elements, placing them in order by both atomic mass and chemical properties, he noticed "gaps" in the data—places where his periodic law suggested there ought to be an element, but none was yet known. Rather than considering this a flaw in his model, however, Mendeleev used the law to predict the existence of three such elements, which he called "eka-boron," "eka-silicon," and "eka-aluminum," after the elements that would be directly above them in the table.* On the basis of his periodic system and the properties of neighboring elements, he made extremely detailed predictions of the properties of these elements. For example, in 1875, he predicted that "eka-aluminum"

---

* The prefix *eka-* comes from the Sanskrit for "one."

should have an atomic weight of 68, a density of 5.9, and an atomic volume of 11.5 and should melt at a "rather low temperature." These predictions are impressively accurate: The atomic weight of gallium is 69.7 units, its density is 5.904, its atomic volume 11.8, and its melting point 29.8°C (85.6°F), rather low for a metal.

Mendeleev's periodic law allowed him to predict this wide range of properties, entirely from gallium's location in his periodic table. Gallium sits just to the right of zinc, directly beneath aluminum and just above indium on the modern periodic table. The predicted weight is 3 units more than zinc (atomic weight 65), and the other properties are similar to those of aluminum and indium—in fact, in another place, Mendeleev explicitly states that several properties "present the average between those of aluminum and indium." He goes on to predict a long list of compounds and reactions of gallium, all on the basis of the analogous compounds and reactions for zinc, aluminum, and indium.*

One of Mendeleev's most eerily accurate predictions is a bit of a cheat on his part, though. He writes that "it is probable that the metal in question will be discovered by spectral analysis, as were indium and thallium." The 1875 paper from which this was quoted was written *after* Mendeleev heard of the discovery of a new element earlier that year by the French chemist Paul Émile Lecoq de Boisbaudran.† The Frenchman

---

* The spot to eka-aluminum's right is occupied by eka-silicon, which had not yet been discovered to serve as a basis for comparison. Otherwise, Mendeleev probably would have drawn on its properties as well.

† In keeping with tradition, de Boisbaudran chose the name for the new element. Thus *gallium*, in tribute to the Latin word for his native France. The word also bears a strong resemblance to *gallus*, Latin for "rooster," which in French is *le coq*, something de Boisbaudran insisted was a coincidence.

had spotted a new spectral line that didn't match the characteristic colors emitted by any known element in some ore samples he was studying. This small bit of post-diction can be forgiven, though, because Mendeleev was dead-on in so many of his other predictions.

The 1875 discovery of gallium was followed in 1879 by the discovery of scandium by Swedish chemist L. F. Nilson. Another Swedish chemist, Per Cleve, noted the similarity between the properties of scandium and Mendeleev's "eka-boron" and published an influential paper placing the predicted and observed properties side by side in a table. In 1886, "eka-silicon" was discovered by Clemens Winkler in Germany (and was named germanium), and again, the striking similarities between Mendeleev's prediction and the observed properties were quickly pointed out by V. F. Richter.

The discovery of the three "eka-elements" and the close correspondence between their properties and Mendeleev's predictions secured Mendeleev's reputation as the father of the periodic table. While other chemists of the day developed their own organizing schemes for the elements and even made some predictions on that basis, none of their schemes were as bold or as detailed as Mendeleev's. His extreme confidence in his classification system gave him a solid basis for new science and earned him a prominent place in the history of science.

FIXING THE TABLE: CHEMISTRY SINCE MENDELEEV

While Mendeleev's periodic law was quickly and enthusiastically adopted by chemists around the word—as Michael Gordin notes, "rarely has a foundational scientific development been introduced with so little debate"—the underlying reasons for the system remained mysterious into the twentieth century. The system worked, and as new elements were discovered, the table easily expanded to accommodate them, including an entirely new column, the noble gases, nearly all of which were discovered after Mendeleev's original work.* The explanation for this periodicity, though, was not worked out until after the structure of atoms was understood.

The modern picture of the atom has a central, positively charged nucleus orbited by negatively charged electrons. These electrons are arranged into "shells" with discrete energies (usually visualized as orbits of increasing radius; see Chapter 8), and each shell can hold a particular number of electrons (in order of increasing energy, the shells can hold 2, 2, 6, 2, 6, 2, 10, and 6 electrons). As you move through the periodic table, the number of electrons in each element increases, and the electrons fill up these shells in order, with each new electron going into the next available state. This process of filling up shells of increasing energy is central to the modern understanding of chemistry.

The periodic properties Mendeleev observed are determined by this arrangement of electrons. Chemical reactions are driven by interactions in which each atom attempts to fill its outermost (that is, highest-energy) shell by stealing electrons from, sharing electrons with, or giving up electrons to other atoms. The structure of these shells repeats over and over (a shell holding 2 electrons is followed by one holding 6), leading to the periodic repetition of these properties; each column of the periodic table represents a family of elements with the same number of electrons in its outermost shell. The alkali metals are highly reactive because they each have in

---

* Helium had been tentatively identified, thanks to a spectral line from the Sun in 1868, but was not discovered on Earth until 1895. The other noble gases (neon, argon, krypton, xenon, and radon) were all discovered in the late 1890s.

their outermost shell a single electron, which is easily given up to other atoms in chemical reactions. Similarly, the halogens are highly reactive because they each have an outermost shell that is nearly full, lacking only a single electron, which they can steal from other atoms. The noble gases get their name because they each have their outermost shell filled to capacity and thus have no need to engage in chemical reactions involving the sharing of electrons.

The modern atomic model also explains the spots where Mendeleev's system appears to break down, such as tellurium and iodine, whose atomic weights are in the "wrong" order. The problem here is that Mendeleev was using the atomic weight to order his table, but weight is only an imperfect proxy for the correct factor, the number of protons present in the nucleus of the atom. The number of protons in an atom is called the *atomic number*. As atoms are usually electrically neutral, the atomic number also determines the number of electrons filling the atom's shells.[†] The atomic weight includes both the number of protons and the number of neutrons in the nucleus and doesn't necessarily increase with atomic number. Mendeleev was correct about the ordering of tellurium (atomic number 52) and iodine (atomic number 53) according to their chemistry, and thus the atomic number, but iodine tends to have fewer neutrons than tellurium and thus a lower average mass.[‡]

[†] The mass of the electron is tiny compared to that of a proton—both protons and neutrons are around 1,837 times heavier than electrons—so the number of electrons does not significantly affect the mass of an atom.

[‡] Technically, both tellurium and iodine have multiple isotopes, with the same number of protons but different numbers of neutrons. Some tellurium isotopes are lighter than some isotopes of iodine, but the average mass of naturally occurring tellurium isotopes is greater than the average mass of naturally occurring iodine isotopes. The average mass is what chemists of Mendeleev's day could measure and was used to order his table.

There is no way for Mendeleev to have known the distinction between atomic mass and atomic number, of course, as only one of these particles, the electron, was discovered before his death in 1907.* The atomic nucleus wasn't discovered until 1909, and the ordering of elements by atomic number was not established until 1913 (by Henry Mosely). The term *proton* was not associated with the particle until 1920, and the neutron, the particle responsible for the distinction between atomic number and atomic mass, was not discovered until 1932, a quarter century after Mendeleev's death and more than sixty years after his work on the periodic table.

The remarkable success of his periodic table in spite of all this is a testament to the power of scientific model-making and the collector's impulse to sort things. Studying only two-thirds of the naturally occurring elements and sorting them by patterns in their properties allowed Mendeleev to anticipate science that would not be discovered for decades, and his scientific collector's instinct gave birth to one of the most useful and iconic symbols of modern science.

CATEGORIZATION IS CRITICAL

Sorting and classification play a key role in many other areas of science. Starting close to chemistry, after the discovery that there are multiple isotopes of the same element, nuclear physicists began to wonder what determined which isotopes are stable enough to be found in nature and what determines the lifetimes of nuclei subject to atomic decay. For example, the heaviest stable isotope known is lead-208 (with 82 protons and

---

* Mendeleev preferred to think of elements as indivisible and was never particularly happy about the idea of electrons as components of atoms.

126 neutrons), but adding just 1 neutron to that nucleus gives lead-209, which decays within a few hours. This difference in stability must result from some internal structure of the nucleus, similar to the electron shells that determine the chemical characteristics of the different columns of Mendeleev's table. But the pattern isn't the same. As the atomic number (and thus the number of electrons) increases, the chemical properties repeat every 8 steps for light elements, or every 18 steps for heavier ones, but the picture for nuclei is more complicated. Chemical properties are determined only by the number of electrons, but nuclei have two components, protons and neutrons, and the number of each component required for a stable isotope doesn't follow the same pattern as the electron shells.

The key to unraveling this puzzle is the classification of the states, which was realized by Maria Goeppert-Mayer in 1949. Stability in nuclei isn't merely a matter of counting the numbers of protons and neutrons—the internal states and interactions of those particles must also be taken into account. In particular, the protons and neutrons have angular momentum associated with their motion inside the nucleus (they're not literally orbiting like planets in the solar system, though the mathematical description is similar), and they have intrinsic spin (again, they're not literally little spinning balls, but the math is similar). The spin and the orbital angular momentum interact with each other, changing the energies of the nuclear states for a given number of protons and neutrons.

Once Goeppert-Mayer realized the importance of this interaction, she could sort the states into a series of *nuclear shells* containing protons or neutrons with the same angular momentum. These shells produce a different set of *magic numbers* than those for electrons in chemistry: The most stable isotopes are those containing 2, 8, 20, 28, 50, 82, or 126 protons or neutrons (the two particles have independent sets of shells). Lead-208 is an exceptionally stable element because it has a magic number of both protons (82) and neutrons (126); adding another neutron moves it away from this "doubly magic" state, and the resulting isotope is unstable. Goeppert-Mayer shared the 1963 Nobel Prize in Physics with Hans Jensen, who independently worked out the same model, again by finding the appropriate way to classify the states of the particles inside the nucleus.

Of course, the science most associated with classification is biology, with different organisms assigned Latin names according to a hierarchy of categories. High school students in my day were expected to memorize "Kingdom, Phylum, Class, Order, Family, Genus, Species," the names of the categories from broadest to most specific. This system has its origin in the work of the Swedish biologist Carl Linnaeus in the 1730s and is at the root of a lot of the teasing directed at biologists by physicists. In addition to Rutherford's stamp-collecting gibe a few chapters back, there's Enrico Fermi's famous reply when he was corrected by a student after mixing up two subatomic particles: "Young man, if I could remember the names of all these particles, I would've been a botanist!"

While biological arguments over whether some tiny difference between two organisms really demands the identification of a new species can seem interminable to outsiders and even to some biologists (many modern scientists prefer a different classification system), the taxonomy introduced by Linnaeus was critical to Darwin's development of the theory of evolution. Arguments about what, exactly, constitutes a new species, and especially the question of whether species were fixed and immutable (as many Victorian scientists wanted to believe for religious reasons) or could change over time helped inspire Darwin's work. The taxonomic system groups together species with similar characteristics and thus is easily adapted to a sort of evolutionary family tree showing interrelationships between species that share common ancestors.

Sorting and classification can be essential in nonscientific work, as well. If you've ever needed to write a document with more depth than a shopping list, you've probably faced a version of Mendeleev's original dilemma: needing to present a large amount of information about some subject in an efficient manner. This challenge crops up regularly in education at all levels—from the college classes I teach, to the sixth-grade lessons my father taught before his retirement, to the kindergarten curriculum my daughter is now going through. And it occurs in engineering, business, law, politics, medicine, and any other profession that requires you to convey the details of some situation to a person who isn't already familiar with it.

Given a sufficiently large mass of facts to be conveyed, this can seem an impossible task, and "I don't know how to start" is one of the most common complaints from students assigned research papers. The answer is often to start by trying to classify the various facts and sort them into some sensible order. Many times the process of dividing information into categories will make patterns become apparent, like Mendeleev's periodic law, and a natural order will emerge. Once that happens, all that remains is putting words on paper.

In fact, the book you're reading is the result of just this sort of process. While I had the idea of a book about scientific thinking in everyday activities for some time and had illustrated it with a few anecdotes about the history of science, I was having difficulty arranging my ideas into a coherent project. Then my agent, Erin Hosier, pointed out that the look, think, test, tell process of science provides a nice way to sort and classify stories, at which point everything fell nicely into place. The structure of the book became clear, and numerous other stories and hobbies suggested themselves as illustrating one aspect or another. The research and writing wasn't exactly easy, but once I knew the structure, the path forward was always clear.

As strange as some of their shelving schemes may seem, then, bibliophiles and record collectors are using their inner scientists. Sorting large amounts of stuff in many different ways can lead to great results. Sometimes the result is an aesthetically pleasing set of bookshelves, sometimes it's a coherent and convincing legal brief, and sometimes it's the periodic table of the elements.

Chapter 6

ASKING THE
ALLOWED QUESTIONS
BRIDGE AND SCIENTIFIC THINKING

My father, now retired after thirty-plus years as a sixth-grade teacher, recently took up playing bridge. This puts him in company with millions of people around the world who play. There are bridge clubs that meet to play in person, countless websites where you can play online with others, and numerous bridge applications for playing on your computer or smartphone. It's one of the most popular card games in the world, and also a nice demonstration of scientific thinking in action.

The card play in bridge is similar to other trick-taking games: A game generally requires four players, in teams of two, with the partners seated across from each other. When the first card in a trick is played, the other players must play a card of the same suit, if they have one, and the highest card in that suit wins the trick. A player who doesn't have a card in the suit that was led is free to play any other card. In each hand, one suit is designated the trump suit, and the lowest card of the trump suit beats the highest card of any other suit. At the end of a trick, whichever player played the highest card of the suit that was led, or the highest trump card, wins the trick and gets to play the first card of the next trick.

What sets bridge apart from other trick-taking games, like hearts or spades, is the bidding process by which the trump suit is determined. At the start of each hand, players take turns bidding, according to the cards in their hand. Each bid is a number from one to seven, followed by either a suit or "no trump" (so, for example, "one spade" or "three no trump"), and players take turns either increasing the previous bid or passing. Once one player has made a bid that none of the other players is willing to exceed, both the trump suit and the goal of the game are set: The trump suit is whatever suit was last bid, and the player who made the last bid and his or her partner must win at least six plus the number of the last bid tricks to win that hand. So, for example, a final bid of "three clubs" means that clubs will be the trump suit for the hand, and the pair including the final bidder must win at least nine (six plus the three from the bid) of the thirteen tricks to win the hand. A bid of "seven no trump" is the ultimate in confidence (or arrogance): The partners who win the bid must win all thirteen tricks of the hand, with no trump suit.

Since the ultimate success or failure of a hand depends on the number of tricks taken by *both* partners, it would obviously be very helpful to know exactly what cards your partner holds before making a bid. The rules of the game forbid directly exchanging this information, though. Instead, the contents of a hand must be conveyed through the bidding process itself. A Byzantine set of conventions has grown up around bidding, determining exactly what you should bid when and how to interpret the bids made by your partner. Done well, though, the bidding process can be remarkably effective—experienced players will enter a hand knowing almost exactly what cards are held by the other players.

Bidding in bridge is what makes taking up the game an intimidating prospect for a new player, but it's one of the main attractions of the game for those who are passionate about it. It's also very much like the process of doing science.

Bidding is an attempt to determine specific and detailed information about another player's cards using a limited and formalized set of questions and responses. This is exactly the situation many scientists find themselves in when they approach the natural world: They want to learn some very specific information about how the world operates—the makeup of a subatomic particle, the structure of a molecule, or the effect of a drug. But to do this, they have to piece together a picture of what's really going on from the results of a limited set of possible experimental tests—what happens when a subatomic particle collides with other particles, the way light bounces off a crystal of some substance, whether a test animal lives or dies. The results of these experiments, combined with some background knowledge about how the world works, can indirectly tell us an amazing amount about the workings of the universe, through the same sort of process that bids combined with knowledge of the bidding conventions tell a good bridge player the contents of the partner's hand.

ASTRONOMICAL CONSTRAINTS

No science operates under stricter constraints than astronomy. Even geologists studying events that took place millions of years ago can subject their rock samples to a wide array of chemical and physical tests. But astronomers can look at one and only one thing: the light arriving from distant objects. And yet, given that extremely limited source of information, astronomers have developed an amazing understanding of the universe in which we live, its origin, and its eventual fate.

In Chapter 4, we talked about the most obvious aspect of looking at light from distant objects, which is just looking at the brightness of a given source over time. This process is what allowed Henrietta Leavitt to find the relationship between period and luminosity—a relationship that makes Cepheid variables a useful yardstick for measuring the distances to other stars and galaxies. Similarly, a variation in the intensity of a radio source caught Jocelyn Bell's attention and led to the discovery of pulsars. The slight dimming of a distant star as a planet passes between it and us on Earth has allowed scientists to discover and characterize hundreds of extrasolar planets, including some discovered by citizen scientists using the Planet Hunters website.

Astronomers have a second tool as well: spectroscopy, the study of the range of different wavelengths emitted by a distant object. Spectroscopy provides the basis for the most amazing discoveries in astronomy.

The simplest bit of information that spectroscopy can provide is the temperature of the object emitting the light. All objects emit light with a characteristic spectrum that depends on the temperature—like a piece of metal that glows red, then yellow, then white as it's heated—so measuring the spread of wavelengths tells us the temperature. The Sun has a surface temperature of around 6,000 kelvins (K) and appears yellowish.* The giant star Betelgeuse has a lower temperature, around 3,500 K, and appears reddish, while Sirius, with a temperature of 9,900 K, looks blue-white. The relationship between spectrum and temperature extends well beyond the

---

* The kelvin is a unit of temperature equivalent to 1° Celsius (1°C), but the kelvin scale begins at absolute zero, or −273°C. Room temperature is typically a little under 300K, and water boils at 373K.

visible part of the spectrum: The entire universe is pervaded by the cosmic microwave background radiation, light left over from just a few hundred thousand years after the Big Bang, with a spectrum peaked at a wavelength of about a meter, corresponding to a temperature of 2.7 K.

More importantly, though, individual atoms and molecules also absorb and emit light, but not as a broad spectrum. An atom of a particular element will emit and absorb only at certain very specific wavelengths, with the exact spectrum providing a unique fingerprint for the element in question. Measuring the characteristic pattern of wavelengths associated with particular elements allows astronomers to deduce the composition of distant objects from the spectrum of light that the objects emit. Spectroscopy has even led to the discovery of new elements. For example, helium takes its name from Helios, the Greek god of the Sun, because it was first discovered as a bright line in the solar spectrum in 1868. Because helium is very rare in the Earth's atmosphere, the element wasn't isolated in the laboratory until 1895, almost thirty years after its discovery.

Knowing the temperature and composition of stars many light-years from Earth is impressive enough, but spectroscopy provides more than that. Careful measurements of the wavelengths of light from distant stars and galaxies often show the pattern characteristic of a particular element, but not *quite* at the wavelength you would expect to see on Earth. The difference between the observed and expected wavelengths tells astronomers about the motion of the source.

The change in wavelength is due to the *Doppler shift*, a wave phenomenon discovered by Christian Doppler in 1842. The Doppler shift is responsible for the familiar change in pitch of the sounds from a passing car. As the car approaches, it catches up to the sound waves it has just emitted, producing a shorter wavelength and higher frequency, which we hear as a higher pitch. As the car recedes, it "runs away" from the previous waves, leading to a lower pitch. The "eeeeeeee-owwwwww" noise little kids make to indicate a car zooming past represents the noise we hear as the car goes by and changes from approaching to receding.*

While the Doppler effect is most familiar with sound, it works for any kind of waves, including light. The spectrum of light emitted by a star that's moving toward us will be shifted to shorter wavelengths (called a *blueshift* because blue is at the short-wavelength end of the visible spectrum), and the spectrum emitted by a star moving away from us will be shifted toward longer wavelengths (a *redshift* because red is at the long-wavelength end of the spectrum). The size of the shift depends on the speed of the motion, so astronomers looking at the spectrum of light from a star can deduce both what direction the star is moving, and how fast. Edwin Hubble used the Doppler shift of light from distant galaxies to show that, in general, galaxies appear to be moving away from us, and the farther away a galaxy is, the faster it's moving. This relationship between distance and apparent speed results from a general expansion of the universe and is the crucial bit of evidence pointing toward the Big Bang theory of cosmology.

Astronomers are almost as tightly constrained in the questions they can ask about the universe as bridge players bidding on a new hand, with only two possibilities: "How much light does this emit?" and "What wavelengths of light does this emit?" Like the conventions employed by bridge players,

---

* The Doppler effect was first demonstrated in 1845 by C. H. D. Buys Ballot using a brass band on a railway car and trained musicians by the tracks listening for the change in pitch.

though, knowledge of the underlying physics of thermal radiation, atomic emission lines, and the Doppler effect allows astronomers to interpret the answers to those limited questions to extract a huge amount of information. Put together with Newton's laws of motion and gravity, a study of the light emitted by distant objects even reveals the presence of things that *don't* emit any light, which has forced a dramatic reconsideration of the makeup of our universe.

GALAXY ROTATION AND DARK MATTER

Vera Rubin has spent more than sixty years studying galaxies, in one way or another. Her career trajectory has been somewhat unusual, largely because of the difficulty of being a woman in a mostly male profession.[†] Thanks to a strong personality and a supportive family, however, she carved out a distinguished career for herself and broke a path for many other women.[‡] Along the way, she helped revolutionize our understanding of the universe and our place in it. We now know that the "ordinary" matter that makes up everything we interact with on Earth accounts for only about 4 percent of the

---

[†] Even today, only around 19 percent of astronomy faculty are women, but Rubin began her career at a time not far from the days when Henrietta Leavitt was paid twenty-five cents an hour and had her work published under someone else's name.

[‡] Her biography includes an impressive list of firsts, including being the first woman invited to observe at the famous Mount Wilson Observatory, at the time the largest telescope in the world. The application form she was sent with the invitation had the sentence "We are unable to accept applications from women" at the top, with the word "usually" added in pencil (Robert J. Rubin, "Vera Cooper Rubin, 1928–," in *Out of the Shadows: Contributions of Twentieth-Century Women to Physics*, ed. Nina Byers and Gary Williams [New York: Cambridge University Press, 2006]).

universe. The vast majority of the universe is made up of stuff we can't see, and Vera Rubin is a big part of how we know that.

From the start, her work has revolved around the use of spectroscopy to study the behavior of galaxies. The first project she worked on, as a master's student at Cornell, aimed to test whether the universe as a whole was rotating, an idea promoted by the physicist George Gamow and the mathematician Kurt Gödel. This is a perfectly valid mathematical solution to Einstein's equations of general relativity, the same equations that predict the expansion of the universe detected by Hubble using the Doppler shift of galaxies.

If the universe is rotating, carrying galaxies along with it, the rotation should be reflected in the spectrum of light from galaxies. The Doppler shift can only measure the velocity along the line of sight, but unless we happen to be exactly at the center of any rotation (which is vanishingly unlikely), the rotational motion would necessarily mean that some galaxies are moving toward us, and others are moving away. The difference would show up as a *peculiar velocity*, a deviation from the general pattern identified by Hubble. Rubin used existing measurements of the Doppler shifts of 108 galaxies to determine their velocities and subtracted from these velocities the velocity you would expect from Hubble's law, given their distance from us. Then she looked for any pattern to the distribution of peculiar velocities.

She presented these results at the 1950 meeting of the American Astronomical Society, to a rather hostile reception.[*]

---

[*] Rubin was accompanied by her husband, parents, and one-month-old son at the meeting. A senior professor in her department had suggested that he present the work instead, under his name, because she obviously couldn't go with a small child. She replied, "No, I'll go," and found a way to make it work.

Many astronomers took for granted the idea that the universe was perfectly uniform, and they thought it ridiculous to even look for patterns that would deviate from uniform expansion. In the years since, though, astronomers have realized that galaxies are not distributed uniformly, but are pulled together by gravity into clusters and superclusters. Techniques very similar to those used by Rubin in 1950 have subsequently been used to study the structure and evolution of the universe on a vast scale, and studying the motion of large clumps of galaxies has grown into a respectable subfield.

Maybe the most important result of her 1950 presentation, though, was that it brought her to the attention of George Gamow, who with others proposed the idea of a rotating universe. Rubin had moved to the Washington, D.C., area with her husband Bob and their son. By chance, Bob shared an office at the Johns Hopkins Advanced Physics Laboratory with Ralph Alpher, a former student of Gamow's.[†] When Bob told Alpher about his wife's master's thesis project, Alpher relayed it to Gamow, and one night, Vera Rubin was surprised to get a phone call at home from Gamow asking her deep questions about the motion of galaxies.

These discussions helped reconfirm Rubin's desire to be a professional astronomer, and she began her Ph.D. studies

---

[†] Alpher worked with Gamow on a number of problems in astrophysics, including the first prediction of the cosmic microwave background radiation and a seminal paper on the creation of elements in the aftermath of the Big Bang. Gamow was infamous for his quirky sense of humor, and as a joke inserted the name of the eminent physicist Hans Bethe (without the knowledge of either Alpher or Bethe), so the author list would read "Alpher, Bethe, Gamow" sounding like the first three letters of the Greek alphabet. In fact, the paper is often referred to as the "α, β, γ" paper and is an important milestone in the history of cosmology.

at Georgetown University, with Gamow as her advisor.[*] She continued to do distinguished work and received her Ph.D. in 1954, despite considerable personal inconvenience.[†] After receiving her degree, she stayed on at Georgetown for another ten years, conducting research and teaching, before moving to the Department of Terrestrial Magnetism (DTM) at the Carnegie Institution in Washington, D.C., in 1965.

Rubin's most important contribution to astronomy began during her time at Georgetown and concerns not the motion of galaxies, but motion *within* galaxies. Her interest started with a class project at Georgetown in 1962, where she and her students used the Doppler shifts of stars within the Milky Way to calculate the stars' velocities and thus determine how fast they were moving in their orbit about the center of the galaxy. Rubin and her students expected that stars farther away from the center would be moving slower, just as the outer planets in the solar system do. Surprisingly, though, the orbital velocity remained nearly the same, all the way out to the farthest stars the team measured.

Rubin had identified an important problem, though nobody realized it at the time; most astronomers wrote it off as a glitch, and the paper that resulted from the class project was mostly ignored.[‡] The puzzle of these too-high velocities would return, however, after she moved to the DTM and began to

---

[*] She was only out of school for about six months, but found being away from astronomy a miserable experience and was thrilled to get back into it.

[†] At the time, she didn't drive, and they had two small children. So after work, Bob would pick up Vera's mother to watch the children, then drive Vera to the observatory, and eat his dinner in the car while she was in class.

[‡] In a 2000 review article, Rubin drily notes that this paper "apparently influenced no one and was ignored even by the senior author when she returned to the problem of galaxy rotation a decade later."

study the problem that would define her career: the rotation of distant galaxies.

The rotation of a galaxy is, of course, far too slow to watch directly, but the rotation leaves a clear imprint on the light from the galaxy. Light from the central region of the galaxy is Doppler shifted to the red by the expansion of the universe. But for a rotating galaxy, there is an additional Doppler shift that varies across the galaxy. On one side of the core, the stars are moving toward us, and thus their light will be shifted slightly to the blue (resulting in an overall shift that's smaller than expected), while on the other side, stars are moving away, and their light is shifted further to the red.

The light from distant galaxies is generally very faint, and spreading it out to take a spectrum makes it even fainter. The first measurement of the rotation of a galaxy was made in 1916, for the famous Andromeda galaxy. The task required eighty-four hours of telescope time, spread out over weeks, and the careful reuse of the photographic plates. Even then, only the brightest central region of the galaxy provided a useful spectrum. The study of galaxy rotation was picked up by Margaret and Geoffrey Burbidge in the 1950s and 1960s, but the required exposures were still long and difficult. Fortunately, around the time Rubin joined the DTM in 1965, her new colleague Kenneth Ford invented a detector that reduced the required time to just a few hours, enabling multiple measurements in a single night of observing. More importantly, Ford's detector allowed Rubin and Ford to obtain useful spectra farther from the center.

Rubin and Ford began their observations in 1967 and 1968, collecting spectra from sixty-seven star-forming regions in Andromeda and calculating the velocity of those regions. When they made a graph showing the rotation speed of those regions as a function of the regions' distance from the center, the two researchers saw the same sort of behavior Rubin and her students had seen with the Milky Way: Rather than dropping off as expected, the speed of these regions remained the same, all the way to the outermost edge of the visible galaxy. Rubin and Ford repeated the measurement for many other galaxies through the 1970s and found the same thing, every time.

Figure 6.1 illustrates the process of determining the rotation of the spiral galaxy NGC 2090 (a visible-light image is on the top left). The top right portion of the figure shows a portion of the spectrum, around a wavelength of 650 nanometers.* The large, dark, horizontal bar is the central region of the galaxy, where there are lots of stars producing a huge amount of light; above and below this, the light from the upper and lower arms of the galaxy is spread out with longer wavelengths on the right, and shorter on the left. The spectrum shows two types of interesting features: straight, vertical lines made by light from atoms in Earth's atmosphere, and three curved lines with a characteristic S shape (one marked with a black arrow, one to the right of that, and a very faint one to the left). The three curved lines are produced by atoms in hot clouds of gas in NGC 2090 where new stars are being formed.

The two features marked by arrows are light of a particular red wavelength emitted by hydrogen atoms: the straight line on the left (marked by a white arrow) from atoms in Earth's atmosphere, and the curved line on the right (marked by a black arrow) from NGC 2090. The difference between the

---

* This spectrum was obtained by my colleague Rebecca Koopmann in the mid-1990s; she used CCD cameras on the 4-meter telescope at the Cerro Tololo Observatory in Chile and required only about thirty minutes. Astronomy has come a long way since the 1960s.

Figure 6.1. *Bottom:* Plot of the rotation speed of the galaxy NGC 2090 (*top left*) as deduced from its spectrum (*top right*).

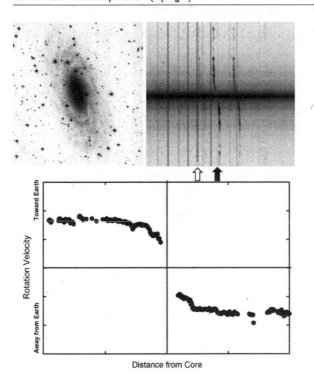

Source: Image and spectrum from the Cerro Tololo Inter-American Observatory, provided by Rebecca Koopmann. Used with permission.

positions of the arrows reflects the Doppler shift due to the expansion of the universe; NGC 2090 is moving away from us at 921 kilometers per second. The S-curve is caused by the rotation of the galaxy itself—stars on the top of the image are moving toward us, so their light is shifted to the left (blue), while stars at the bottom are moving away, and their light is shifted to the right (red). If you trace the line from top to bottom, you can imagine it like the sound of a passing police car: as it approaches, the pitch is shifted up, and as it recedes, the pitch is shifted down; in the middle, there's a rapid change from one to the other. The difference between the two sides of the S-curve indicates a rotation velocity of around 160 kilometers per second. The large graph at the bottom shows the velocity inferred from the Doppler shift. Strikingly, outside of a small region near the center of the galaxy, the rotation speed is nearly constant, out as far as there is visible light to measure.

Every galaxy Rubin and Ford measured showed the same pattern. Other astronomers began looking at the same problem. Among them was Mort Roberts, who confirmed Rubin and Ford's observations by looking at the radio emission of gas clouds in Andromeda and twenty-two other galaxies. Roberts found that the velocity remained constant even for gas clouds well beyond the visible disk of the galaxies.

One or two such observations might be written off as a measurement error, but the accumulating mass of evidence through the 1970s and 1980s, observed by many astronomers at different wavelengths, could not be ignored. Something was very, very wrong with the conventional idea of what galaxies are made of.

EVIDENCE OF THINGS UNSEEN

The galaxy spectra collected by Rubin and Ford were the limited information they could obtain under the constraints of astronomy, like the bidding history in a bridge game. To go beyond the bare facts of the measured frequency shifts requires further knowledge about the conventions of the process, in this case, the laws of physics.

Newton's laws of motion tell us that an object in motion will tend to remain in motion in a straight line at constant speed unless some force acts to change it. Any object moving along a circular path, then—a yo-yo being swung around by a child, a turning car, or a star orbiting the center of a galaxy—must be experiencing a force pulling it toward the center of the circle. The size of the force needed to hold the object on a circular path increases as the speed increases—swinging a yo-yo very rapidly is harder than swinging it more slowly—and decreases as the size of the circle increases. Knowing the speed of the object and the radius of the curve, then, tells you the force it has to be experiencing.

For a yo-yo, the origin of the force is obvious—it's supplied by the string. The car is a little more subtle, but anyone who has tried to steer on slippery roads knows that friction between the tires and the road is the crucial interaction. An orbiting star, not in direct contact with anything else, is held in orbit by the force of gravity pulling it toward the center. The force of gravity depends only on the mass of the two interacting objects and the distance between them, so knowing the speed of an orbiting star toward the outside of a spiral galaxy tells you something about the total mass of the galaxy.

Astronomers can estimate the amount of mass in a galaxy from the amount of light emitted by its stars and gas—the more light you see, the more stuff there must be to emit that light. A spiral galaxy like Andromeda or NGC 2090 extends over a large region of space, but the largest concentration of mass is in the central bulge of the galaxy.* The arms of the galaxy contain a substantial amount of stuff, too, but the density of material drops off as you move out.

---

* Observations in x-ray and radio wavelengths suggest that for a typical galaxy, a significant fraction of the mass in the bulge is concentrated in a supermassive black hole at the very center. The black hole at the center of the Milky Way packs some four million times the mass of the Sun into a region smaller than the orbit of Mercury.

In light of this model, it's reasonable to expect the stars in a galaxy to behave like the planets in the solar system: The stars toward the outside should move more slowly than the stars near the center, just as the outer planets orbit much more slowly than the inner ones. Even though the planet Neptune feels the gravitational pull of not only the Sun, but also all the other planets, the total mass of the planets doesn't add up to enough to overcome the decrease in the force with distance from the Sun.* For this reason, Neptune moves at a fifth of the speed of the Earth, taking some 165 years to complete one orbit.

Putting together the unexpectedly constant rotation speeds measured by Rubin and her colleagues with well-known physics, then, leads to a surprising conclusion: Galaxies must contain a vast amount of mass that we can't see.† If the stars near the outside of a galaxy are moving at the same high speed as the stars closer to the center, the force the outer stars experience must be larger than expected. A larger-than-expected force, though, means that the total mass inside the orbit must be larger than expected, to compensate for the drop in the gravitational attraction as you move away from the central bulge. And since the rotation speed remains high all the way out to the edge of the visible galaxy (and even beyond, as seen in radio telescopes), whatever this extra mass is, it can't emit any light.

---

* The gravitational pull of other planets is not completely negligible, though. In fact, Neptune itself was discovered through its effects on the orbit of Uranus. These effects are tiny, however, and needed nearly seventy years of careful observations to be detected.

† An alternative theory, Modified Newtonian Dynamics (MOND), tries to address the problem by changing the mathematical form of Newton's laws. This model has a small number of passionate supporters, but galaxy dynamics is not the only effect pointing toward the existence of dark matter, and MOND is not widely accepted.

Like a bridge player deducing from her own hand and the bidding history which other player's hand conceals the queen of spades, Vera Rubin put together spectroscopic observations and physics knowledge to deduce the presence of *dark matter* in galaxies. Galaxies are surrounded by vast halos of invisible matter, which accounts for 90 percent of the total mass; the light we see comes from an insignificant fraction of what's really there.

The idea of dark matter is a radical suggestion that took a long time to become accepted, but numerous other lines of evidence point in the same direction. The idea of vast quantities of unseen matter was first suggested by Fritz Zwicky in the 1930s; he coined the term *dark matter* based on observations of galaxies in clusters. Zwicky's theory wasn't widely accepted, but more modern studies of clusters have vindicated his suggestion, and the observed structure of galaxies and galaxy clusters only works if dark matter is included. Measurements of the relative amounts of the lightest elements—hydrogen, its heavier isotope deuterium, helium, and lithium—combined with models of the Big Bang suggest that there must be a vast quantity of exotic matter out there somewhere. Moreover, careful measurements of the temperature of the cosmic microwave background radiation left over from the Big Bang can best be explained by including dark matter in the model.

The most dramatic evidence of dark matter comes from a combination of x-ray and visible-light observations of the Bullet Cluster, where two groups of galaxies are colliding with each other.[‡] The collision has heated and compressed the gas

---

[‡] The formal designation of the Bullet Cluster is 1E 0657-558, which isn't nearly as much fun to say.

between galaxies to the point where it emits x-rays, and observations with the Chandra satellite give a clear picture of the amount and distribution of this gas. The mass of the cluster also bends the light, in accordance with general relativity, distorting the images of more distant galaxies, and the amount of distortion allows astronomers to map out the distribution of mass in the cluster. They find that the mass concentration responsible for bending the light is in a very different location from the hot gas. While the stars and gas of the galaxies make an impressive show for telescopes on Earth, the vast majority of the stuff in the cluster is invisible. Its presence can be deduced by combining constrained observations with conventions of physics, just as in Vera Rubin's observations, and leads to the same conclusion: The vast majority of the mass of the cluster is stuff that we can't see.[*]

BACK ON EARTH

The process of deduction from limited information and from knowledge of background conventions is essential to all of science, even in fields with better access to their objects of study than astronomy. Atomic and nuclear physicists deduced the internal structure of atoms from measurements of what comes out when atoms are bombarded with light or other particles

---

[*] The idea of dark matter in galaxies was well accepted by the late 1990s, thanks in large part to Rubin's observations, but the universe had yet another unseen surprise. Observations of supernovae in distant galaxies showed that the expansion of the universe is, in fact, accelerating, indicating the presence of another mystery substance, dubbed *dark energy*. The story of its discovery follows a very similar process of constraints and conventions, but is too long to explore here. Richard Panek, *The 4 Percent Universe: Dark Matter, Dark Energy, and the Race to Discover the Rest of Reality* (Boston: Houghton Mifflin Harcourt, 2011), provides a fascinating history of both dark matter and dark energy and how we know they exist.

(we'll talk more about this in Chapter 8). Physical chemists such as Rosalind Franklin (mentioned in Chapter 2) deduce the structure of complex organic molecules from the patterns x-rays make when passing through a crystal and from a knowledge of the wave nature of matter.

Limited observations and knowledge of background physics can also be used to reconstruct events in the distant past, even some amazingly improbable events. In 1972, French nuclear engineers noticed that uranium ore from the Oklo mine in Gabon had a smaller concentration of the isotope uranium-235 than normally expected. Careful exploration of the mine and the surrounding area revealed that water seeping into the ore deposit about two billion years ago had created a "natural nuclear reactor," in which most of the uranium-235 atoms were split, just as they are in a nuclear power plant. Investigations of the Oklo deposits provide a detailed history of the operation of the "reactor" and help place limits on exotic theories in which the fundamental constants of nature might change over billions of years.

Even in psychology, where you might expect scientists to be able to directly ask a wider range of questions of their subjects, indirect inference plays a crucial role. It's impossible to know exactly what happens inside the mind of another person. Even when you ask a person directly, all you can get are the person's subjective impressions of what he or she thinks. Clever experimental tricks reveal a wide range of cognitive biases showing that we're often very badly misled about what our own brains are doing.[†] Inferences drawn from simple experiments, and more-sophisticated tools like magnetic

---

[†] Many of these are described in Christopher F. Chabris and Daniel J. Simons, *The Invisible Gorilla: And Other Ways Our Intuitions Deceive Us* (New York: Crown, 2010).

resonance imaging scans, are beginning to allow a more complete understanding of what's really going on.

The range of tests that can be employed is greater in other sciences than in astronomy, but every field makes use of the inference skills employed by bridge players. This process also turns up in solving problems in everyday life.

When I was in graduate school, I rented a cheap room in a house that had some slightly dodgy wiring. In particular, the circuit breaker controlling the power to the kitchen was very easy to trip and regularly needed to be reset. One morning, I woke up to find the power off in the kitchen and a note from one of the other residents. He had gotten up in the middle of the night and discovered that the breaker had tripped, but since he didn't know where the breaker box was, he just wrote me a note and went back to bed.

In addition to being annoyed at his laziness, I realized after a bit of thought that his action (or inaction) ultimately represented a failure to think scientifically. At three in the morning, my housemate was constrained from asking me or the landlord where to find the breaker box, but he gave up without realizing that he could use knowledge of the conventions of home construction to deduce the location and get the lights back on. An ugly object like an electrical panel is unlikely to be in plain view in the public areas of the house (which, of course, is why he couldn't see it in the kitchen or living room). A panel is also unlikely to be located in one of the private bedrooms, where it would be inaccessible to anyone but that one tenant. Thus, the only place it could reasonably be would be in the basement, which was both out of public view and accessible to any resident who might need to reset a tripped breaker. And in fact, this process of reasoning was how *I* had originally located the breaker box, the first time I found the lights out in the kitchen.

We don't necessarily think of this reasoning process as something scientific, any more than we think of card play as scientific. In fact, though, we make use of this kind of reasoning all the time. When we navigate in unfamiliar spaces, we use our knowledge of the conventional organization of houses and public spaces to make our search easier: The bathroom in a strange house tends to be upstairs or toward the back rather than in an area visible from the street, plates and silverware tend to be in kitchen cabinets and drawers rather than stored inside the oven, and so on.

Making inferences from constrained information and general conventions is also an essential social skill. It's what we do every time we ask ourselves, "What could he possibly mean by that?" after a cryptic email or text message—we attempt to work out the significance by combining the limited contents of the message with our background knowledge of the sender. In fact, discovering dark matter is arguably an easier problem than interpreting some of the stuff people put on Facebook. If you can successfully navigate modern social media, you're thinking like a card player, and also a scientist.

Chapter 7

DINOSAURS AND MYSTERIES

mong the several hundred books Kate and I own (this, after a couple rounds of purging books from our collection to free up space for the kids) is a complete set of Rex Stout's Nero Wolfe stories, which we assembled over several years of poking through used book stores. These run to thirty-three novels and thirty-nine short stories, written between 1934 and 1974 and featuring the adventures of Wolfe, a corpulent private detective who refuses to leave his house. The tales are narrated by his right-hand man, Archie Goodwin, who does all the legwork for their cases. The series has also been adapted for film and television several times over the years, most recently in an excellent series on A&E in 2001–2002, with the late Maury Chaykin as Wolfe and Timothy Hutton as Goodwin.

The Wolfe series is notable for the length of its run, but is just one of many multi-novel mystery series.* The mystery genre is one of the most enduring and successful in pop

---

* The series is also notable because none of the principal characters age over those four decades, despite the fact that *A Right to Die* (1964) features the adult son of a character from *Too Many Cooks* (1938). There's evidently some sort of group photograph of Dorian Gray aging in the attic of Wolfe's brownstone.

culture. Mystery novels take up a large chunk of shelf space in a typical bookstore, and it would be tricky to find an hour of the day when there isn't some sort of detective show on television. Mystery plots even turn up in other genres—historical fiction, science fiction, even literary novels. Millions of people read or watch some sort of mysteries for relaxation every day.

Many mystery stories, particularly modern shows focusing on forensics (*CSI* and its various spin-offs, *Bones*, etc.) explicitly feature science.* But the connection between scientific thinking and mystery novels extends well beyond those cases where the lead characters are scientists. The most iconic of fictional detectives, Sherlock Holmes, frequently declares that he is practicing "the science of detection," and his methods have their roots in scientific technique.† Scientific thinking of the sort we're dealing with in this book, though, is an integral part of mystery stories not only for the characters, but also for the audience. Well-done mysteries involve scientific thought not only by the fictional sleuth, but also on the part of the readers or viewers following along at home.

THE GENERIC MYSTERY STORY

While fictional detectives are among the most popular characters in literature, their origin in English literature is relatively recent, often dated to Edgar Allan Poe's "The Murders in the

---

* A development regarded as very much a mixed bag by people in the field. On the one hand, the popularity of forensic science in pop culture drives a lot of students toward classes in chemistry and other related disciplines, which makes many college faculty happy. On the other, these shows have been accused of creating unrealistic expectations of the quality of forensic evidence, making life much harder for police and prosecutors.

† The character of Holmes was based in part on a doctor Arthur Conan Doyle knew in Edinburgh.

Rue Morgue" in 1840 and two other stories with same protagonist, Auguste Dupin. Numerous other fictional sleuths have been created over the years—besides Wolfe and Holmes, there are Hercule Poirot, Miss Marple, Sam Spade, Phillip Marlowe, and so on—but all the stories follow a similar trajectory. The generic detective story is fundamentally a puzzle, and the reader follows along with the detective or detectives attempting to solve it.

The story begins with a crime, generally a murder (some stories are kicked off with a theft or a disappearance, but there is almost inevitably a killing before the end). The police investigating the crime are either baffled or fixated on some incorrect solution. The detective protagonist is brought in, either by the police themselves or by one of the people involved. For example, Sherlock Holmes gets many of his cases from Inspectors Lestrade and Gregson of Scotland Yard, while Nero Wolfe is generally hired by one of the principal suspects. Then, through careful observation of the evidence and the characters involved, the detective tracks down the truth. At the end of the story, the criminal responsible is revealed to both the characters and the reader.

There are innumerable variations on the basic formula—"Golden Age" mysteries, hard-boiled detective stories, police procedurals—and books that play with or even spoof the conventions of the genre.‡ Still, in all of these, the solving of puzzles is the essential core that makes reading or watching mystery stories enjoyable. While most readers are not necessarily trying to solve the crime before the characters (as some old Ellery Queen mysteries explicitly challenge the audience to do), mystery fans generally make an effort to keep track of

---

‡ One of my favorite recent mystery series, the Chet and Bernie books by Spencer Quinn, is narrated by the private eye's easily distracted dog.

the facts of the case and how they might fit together. The most common criticisms of mystery stories are either that the author didn't play fair ("They brought in an evil twin in the last chapter") or that the solution was too obvious ("I figured it out in the first ten minutes"). Both complaints arise from the audience's tracking the story along with the characters and trying to put together a satisfying resolution to the problem.

This reading process is essentially scientific. Like scientists, both the fictional detectives and the audience following along are observing the essential facts of the case and fitting them together into a mental model that incorporates and explains all of the observations. It is this model-making step that sets the great detectives apart from their foils on the police force: Nero Wolfe's friend and rival, Inspector Cramer, is also putting together a theory of the case, but always fails to incorporate some detail. As a result, Cramer always reaches the wrong conclusion and is thus regularly embarrassed by Wolfe. The same process is central to science: The key to scientific progress is finding a model that incorporates an entire array of disparate facts.

While all sciences have their mystery-solving aspects, there is perhaps no better analogy between science and mystery than paleontology. The study of dinosaurs has been captivating children and adults for about as long as there have been detective stories—the first complete scientific descriptions of dinosaurs were in the 1820s—and piecing together how dinosaurs lived and died is one of the most compelling mysteries in modern science.

THE MYSTERY OF THE DINOSAURS

Like a good mystery story, the dinosaur puzzle begins with a death. Lots of them, actually—dinosaurs encompassed thousands of species that flourished on Earth over a span of about 150 million years.* Although the last of the dinosaurs were wiped out sixty-five million years ago (as discussed in Chapter 2), they have left traces behind for scientists to study. Some dinosaurs that died and were covered up by sediment relatively quickly have fossilized over the intervening ages, as chemical reactions replaced their bones with less degradable minerals, and the entombing material slowly turned to rock.† As geological processes of uplift and erosion bring these rocks to the surface, bits of bone get exposed, like a corpse turning up in an odd location to begin a murder mystery. These skeletons, along with other traces such as tracks preserved in soft ground that has since turned to stone, and fossilized dinosaur droppings (*coprolites* is the more socially acceptable name for these), are the physical evidence that scientists have to work with.

The problem facing scientists studying dinosaurs is both more and less difficult than that facing a fictional detective. Unlike a detective studying a recent crime, today's scientists have no hope of finding witnesses or corroborating evidence from contemporaries of the deceased. And the mystery that paleontologists are trying to solve is far more extensive than merely determining what killed off the dinosaurs.‡ Scientists want to understand how these creatures behaved when they

---

* Many popular treatments indiscriminately lump any extinct creature of roughly the right age into the category of dinosaurs, but the scientific meaning refers to a fairly specific group of creatures with particular anatomical characteristics. The dinosaur age also lasted an extremely long time—there was more time between the heyday of *Stegosaurus* (150 million years ago) and that of *Tyrannosaurus rex* (65 million years ago) than between *T. rex* and today.

† Some details of this process remain mysterious, which isn't entirely surprising, given the time scale involved.

‡ Though some progress has been made on this, as described in Chapter 2.

were alive. Deducing answers to this question from the limited evidence preserved in ancient rocks is a challenge that even the most accomplished fictional sleuth would find daunting.

At the same time, though, scientists have a far greater quantity of evidence to work with. They're not trying to find the killer responsible for a single death, but are trying to understand the victims of millions of deaths. Somewhat paradoxically, the grand scale of the problem makes it easier to attack—while no single skeleton can provide a complete answer to the mystery, fossils of nearly one thousand different species of dinosaurs have been identified, some species with hundreds of partial skeletons. What's more, thousands of dinosaur trackways are known. All of these traces contribute bits of evidence toward the overall picture.

The science of dinosaurs has changed an incredible amount from the days when I was a child fascinated by the skeletons at the American Museum of Natural History in New York. Back then, I learned much of the conventional wisdom of the day: that dinosaurs were ponderous, cold-blooded creatures akin to modern reptiles. They were believed to be slow-moving, tail-dragging imbeciles (with brains "the size of a walnut" to quote a famous *Far Side* cartoon). The biggest of them were sometimes claimed to live exclusively in the water, because otherwise their bodies couldn't support their vast bulk. Beyond that, many aspects of these animals, such as the appearance and coloration of living dinosaurs, were said to be unknowable.

All of these beliefs have been called into question, and many proved false, in the same way that the too-simple theories advanced by the police in a detective novel are supplanted by a better model pieced together by painstaking scientific detective work. The modern picture of dinosaurs is vastly different, showing them as active ancestors of modern birds, with fascinating traits unlike almost any modern animals. We even have hints about their outward appearance, down to their color, all deduced from traces in ancient rocks.

A STUDY IN SEDIMENT

"I'll tell you one thing which may help you in the case," [Holmes] continued, turning to the two detectives. "There has been murder done, and the murderer was a man. He was more than six feet high, was in the prime of life, had small feet for his height, wore coarse, square-toed boots and smoked a Trichinopoly cigar. He came here with his victim in a four-wheeled cab, which was drawn by a horse with three old shoes and one new one on his off fore leg. In all probability the murderer had a florid face, and the fin-ger-nails of his right hand were remarkably long. These are only a few indications, but they may assist you."

—Arthur Conan Doyle, *A Study in Scarlet*

Our picture of dinosaurs has advanced greatly in almost every area since the late 1970s, when I first encountered them, and there are many excellent articles and books describing recent advances.* As I don't have a full book to devote to the subject,

---

* Brian Switek's *My Beloved Brontosaurus: On the Road with Old Bones, New Science, and Our Favorite Dinosaurs* (New York: Scientific American/Farrar, Straus and Giroux, 2013) mixes great discussions of modern paleontology with personal travelogue, describing how dinosaur science has changed since he was a dinosaur-obsessed boy. Scott Sampson's *Dinosaur Odyssey: Fossil Threads in the Web of Life* (Berkeley: University of California Press, 2009) is more technical, providing a lot of detail about what we know of dinosaurs and how they fit into the larger ecosystem of their time.

I'll focus on just one aspect of the changes, namely, the outward appearance of dinosaurs.

The preceding Sherlock Holmes quote portrays one of his more astounding bits of deduction. The detective gives the police a detailed description of the killer after only a short time spent inspecting the scene of a mysterious murder in an abandoned London house. There are some clear parallels between Holmes's methods in this story and the techniques used by paleontologists studying dinosaur appearance. His deductions are amazing, but they pale in comparison with those that have completely reshaped our view of dinosaurs.

Several of Holmes's deductions—the killer's height, style of shoes, age, and mode of transportation—are based on the detective's study of the tracks left at the scene of the crime.* The stride length indicates that the killer was both tall and young, and the tracks of the cab he came in ultimately provide the key to nabbing him. Similarly, one of the most striking changes in the appearance of dinosaurs over the years has come in part from looking at their tracks.

Thousands of dinosaur trackways are known, fossilized footprints left behind in soft earth that later turned to stone. These cover a huge range of species, spanning more or less the full 150-million-year period when dinosaurs were the dominant large animals on Earth. The trackways include those made by single animals and other trails left by large groups— at least twenty-three dinosaurs left footprints in one trackway in Texas. Some trackways even include evidence of fleeting

---

* In fact, quite a few Sherlock Holmes stories are resolved by Holmes's noticing and following tracks of one sort or another, making a modern reader wonder what, if anything, the Scotland Yard detectives were doing to earn their pay.

interactions, such as a set of tracks in Africa, where one dinosaur appears to have jostled another, making it break stride. One of the most notable features for our understanding of dinosaur appearance, however, is what these trackways do *not* contain, namely, any trace of a dragging tail.[†]

Early reconstructions of dinosaurs all show their tails on the ground. Long-tailed sauropods like *Apatosaurus* have tails that droop limply to the ground, while bipedal theropods like *Tyrannosaurus rex* stand almost completely upright, with the tail propping up the hindquarters (see Figure 7.1, top).[‡] If these depictions were accurate, we would expect to see drag marks from the tails alongside the many footprints these animals left behind. The absence of tail marks indicates that dinosaurs did not, in fact, drag their tails along the ground. Rather, the tail was held stiffly out, nearly horizontal, as a counterbalance to the forward-leaning bodies of bipedal dinosaurs like *Tyrannosaurus* (see bottom of Figure 7.1).[§]

Other deductions that Holmes made are based on close examination of small traces the killer left behind—the long fingernails are suggested by the scratches he left when writing in blood on the wall, and his brand of cigars was deduced from the character of the ash left at the scene. Similar faint traces have led scientists to the most radical revision of dinosaurs in recent years: the idea that many, even most, dinosaurs were covered with fuzz and even feathers.

[†] A fitting parallel to another famous Holmes case, which turns on a dog that didn't bark.

[‡] Conveniently for cheap special-effects artists, this look is easily approximated by a human in a rubber suit—think Godzilla or Barney the Dinosaur.

[§] The *T. rex* in the *Jurassic Park* movies is a reasonably accurate match for the way scientists think the animal looked.

Figure 7.1. *Tyrannosaurus rex. Top:* First attempted reconstruction, from 1905.
Bottom: Modern reconstruction, from 2006.

Source: "Tyrannosaurus skeleton," *Wikipedia*, last updated April 13, 2014, http://en.wikipedia
.org/wiki/File:Tyrannosaurus_skeleton.jpg; and "Tyrannosaurus rex 1," *Wikipedia*, last
updated September 18, 2007, http://en.wikipedia.org/wiki/File:Tyranosaurus_rex_1.svg.

When I was a kid learning about dinosaurs, most books
said that any reconstruction of dinosaur appearance would always be speculative, because only their bones were preserved as
fossils. As more and more fossils have been unearthed, though,
numerous impressions of softer tissues, including traces of dinosaur skin, have been found. While most of these impressions
look scaly, some fossils preserved in particularly fine-grained
sediments show unmistakable imprints of feathers.

This discovery is not as surprising as it might seem, as dinosaurs are now known to be the direct ancestors of modern

birds. This relationship was in dispute for a long time, because of a misunderstanding worthy of Holmes's regular foil Inspector Lestrade: dinosaur skeletons supposedly did not contain a particular type of bone called a clavicle, which is characteristic of birds.* Dinosaurs were thus assumed to be cold-blooded relatives of modern reptiles. In fact, however, clavicles have been identified in numerous dinosaur skeletons, and since the 1970s, the idea that birds are descended from dinosaurs has been widely accepted.

If dinosaurs are the ancestors of birds, at some point the ancient animals must have evolved feathers. Evidence of either feathers or proto-feathers has recently been found in several species from different families of dinosaurs, which suggests that most dinosaurs may have had some sort of fuzzy coating. This observation also calls into question the long-standing belief that dinosaurs were cold-blooded, like modern reptiles. While it's not an absolute indicator of warm-blooded status, the need for the insulation provided by fuzz and feathers suggests that dinosaurs had a hotter-running metabolism than previously believed. This idea, and the connection to birds, is further supported by the discovery of a fossil of a feathered dinosaur perched atop a nest full of eggs, suggesting a parent warming eggs with its body like a modern bird.†

---

* Not the fictitious "intracostal clavicle" that Cary Grant's character is looking for in the movie *Bringing Up Baby*.

† The question of dinosaur metabolism remains somewhat ambiguous, though. In addition to the feathers, growth rings in bones show that dinosaurs grew to immense size over only a few decades, suggesting a higher metabolism than reptiles. Those very growth rings are more common in cold-blooded creatures, however, and a warm-blooded creature the size of a dinosaur could have severe problems with overheating. A final resolution will require new evidence.

The last and most amazing of Holmes's deductions was his correct prediction of the killer's complexion merely from studying the scene of the crime. In *A Study in Scarlet*, this deduction turns out to be an inspired guess based on the amount of blood left at the scene. While paleontologists don't have any dinosaur blood on which to base deductions, they have deduced the color of some feathered dinosaurs using another signature Holmes technique, the study of the very small features.

One of Holmes's iconic props is a magnifying glass, used to study minute details of the crime scene. For dinosaur feathers, a simple magnifying glass is insufficient, but modern science provides a much more powerful tool, the electron microscope. In 2010, two teams of scientists working with fossils from China used electron microscopes to identify tiny structures called *melanosomes* in fossilized feathers. In modern birds, these structures contain the pigments responsible for coloring the feathers. By comparing the shape and distribution of the fossil melanosomes to those of modern birds, one of the two teams reconstructed the full appearance of an extinct feathered dinosaur: glossy black, with white bands on the wings and a red crest on its head. This technique required a bunch of factors to come together in just the right way, but knowing exactly what an extinct dinosaur looked like is one of the most incredible scientific discoveries of the last forty years.

The discoveries described here barely scratch the surface of the new science of dinosaurs, but they should suffice to give some idea of the essentially mystery-like process used to determine everything we know about the lives of dinosaurs. Like great detectives—or practiced mystery readers following along with great detectives—scientists piece together tiny hints to develop a surprisingly detailed picture of creatures that were wiped out sixty-five million years ago.

MYSTERIES OF SCIENCE

Almost any story in the history of science can be cast in the form of a mystery story: Some strange phenomenon is observed, the people investigating it are initially baffled, but eventually a scientist with a keen eye for small details comes along and pieces together the clues to come up with a correct theory. Most of the stories related in this book are mysteries when seen from the right angle.

The resolution of scientific mysteries takes lots of forms, depending on the details of the problem under consideration. The solar system has provided the stage for two great mystery tales with similar beginnings—a planet was observed not *quite* at the place it was expected to be—but very different endings.

In 1846, the planet Uranus had completed nearly one full orbit since its discovery by William Herschel in 1781, but was deviating somewhat from the path expected from the well-known laws governing planetary motion. The French astronomer Urbain Le Verrier looked into the problem and realized that he could explain the discrepancy if Uranus was experiencing an extra gravitational pull from an as-yet undiscovered planet even further from the Sun. Le Verrier worked out roughly where this planet ought to be found—a complicated calculation even today, but a very impressive achievement a century or so before the invention of the computer. Then, on September 24, 1846, two astronomers in Berlin, Johann Gottfried Galle and Heinrich Louis d'Arrest, found the planet almost exactly where Le Verrier had predicted. The mystery of Uranus's orbit was thus resolved by the prediction and subsequent discovery of a new planet.

Some seventy years later, a similar mystery was solved in a very different manner. The planet Mercury, orbiting closest to the Sun, was also deviating from its expected orbit in a puzzling way. Le Verrier was one of the first to notice this shift as well, and following on his success with Neptune, many astronomers tried to explain the anomaly as the effect of an unknown planet orbiting even closer to the Sun. None of these explanations ever held up—the numbers didn't quite work, and the supposed extra planet stubbornly refused to show up in telescopes.

The solution turned out not to be a new planet obeying existing laws, but a radical modification of those laws: Einstein's general theory of relativity, explaining gravity in terms of a curvature of space-time. For most everyday situations, the predictions of general relativity are nearly identical to those of Newtonian gravity, but as the masses involved become larger and the distances smaller, the two theories begin to differ. Mercury is just close enough to the Sun for these differences to become significant. One of the first things Einstein did after completing the theory in the fall of 1915 was to calculate the resulting shift of the orbit of Mercury. Carrying out the calculation, he got an answer that perfectly matched the observed deviation. When he saw this result, he knew his theory was correct, and he became so excited he had heart palpitations.

Of course, the ultimate example of science's solving a mystery may be the Archimedes story from which this book takes its title. The tale has all the elements of a good mystery: a seemingly perfect crime unraveled by a world-class genius. Most crucially, it turns on a small detail, namely, the water slopping over the edge of the tub as Archimedes lowered himself in. Noticing this provided the key insight that led to the method used to foil the thieving goldsmith.

Those fine details are the key to both great science and great detective work. Every detective, from Poe's Auguste Dupin on down to the foul-mouthed homicide cops on *The*

Wire, has relied on picking up tiny details and fitting them into a bigger picture. And this carries over to the audience for those fictional sleuths.

One of the few current television shows Kate and I still make an effort to watch (after the kids go to bed) is the Sherlock Holmes "reboot" *Elementary*, with Jonny Lee Miller as Holmes and Lucy Liu as Watson. The show moves the characters to present-day New York, with Captain Gregson of the NYPD bringing in the eccentric Holmes as a consultant for the department's most difficult cases. Of course, a hundred-odd years after the Arthur Conan Doyle stories, the straightforward methods of detection employed by the original are no longer as dramatically effective, so the modern stories involve multiple false leads and twist endings.

Like most viewers, I try to follow along, but I'm not often successful at guessing the culprit, given all the false leads and reversals. One of the few times I was, the conclusion turned on a tiny detail. About halfway through a first-season episode, Holmes notices the word NOVOCAINE written in block capitals on a piece of paper. The handwriting rules out the second of the investigators' incorrect guesses at the killer, but doesn't seem to point to anybody else. This allowed me to feel superior for a change, because the block letters had identified the killer for me. One of the secondary characters associated with the crime was working a crossword puzzle when first introduced and spent a later scene discussing crosswords with Holmes. Crossword puzzles (which we'll talk more about in Chapter 8) often involve odd words, and people working a puzzle usually fill it in with block capital letters. Given this tiny detail (and the metafictional constraint that the killer was almost certainly going to be one of the actors with a speaking part), I knew the crossword-puzzle solver had to be involved in the crime. I was pleased to be proved right by the end, when Holmes correctly identified the odd word as a crossword answer.

This process of deduction from tiny clues is fundamentally the same thing paleontologists do when they reconstruct dinosaur appearance from faint impressions and electron microscope images of fossilized pigment molecules. The practical lesson to take from both dinosaur scientists and fictional detectives is simply to be mindful of tiny details.* They're often the key to solving a problem, whether it is an exotic question like "What did a creature that died millions of years ago look like?" or "Who was the killer on this TV show?" or a more mundane issue like "Where did we leave the two-year-old's favorite stuffed animal?" If you're an attentive reader or watcher of detective stories, you already have some practice with noticing and recalling the details that make a great scientist or a great detective.

---

* And while the uncanny abilities of most fictional detectives owe a lot to the author's stacking the deck in their favor, with a bit of effort you can come surprisingly close to matching some of these feats. Maria Konnikova's *Mastermind: How to Think Like Sherlock Holmes* (New York: Viking, 2013) offers a detailed look at some of the tricks Holmes uses and the underlying science of how our brains work.

STEP THREE

TESTING

Aristotle maintained that women have fewer teeth than men; although he was twice married, it never occurred to him to verify this statement by examining his wives' mouths.

—Bertrand Russell, *Impact of Science on Society*

Test everything. Hold on to the good.

—1 Thessalonians 5:21

Once a scientific model has been developed, the crucial next step in the process is to test that model with further observations or experiments. This is the crucial step that sets science apart from various flavors of mysticism—alchemy, superstition, religion, and the like.

The testing step was one of the latest additions to the formal process of institutional science. It wasn't until the 1500s that European natural philosophers began to consistently and systematically rely on experiment and observation to distinguish between models. Before that, many arguments about topics we now consider scientific were conducted on the basis of abstract principles. This approach led to a number of situations that strike modern readers as completely absurd. For example, prior to the work of Andreas Vesalius in the 1540s, most of the conventional wisdom about human anatomy was based on the work of the Roman physician Galen around 200 AD. For religious reasons, Galen was unable to dissect human cadavers, and so his treatise on human anatomy was based on studies of pigs and apes and thus incorrectly describes several features of the human body.

It is, however, an oversimplification to say that the testing of models began with the Renaissance, because the general process of science has been with us since the earliest days of our species. The Renaissance merely saw the systematic, institutional application of an informal process that had existed for millennia—humans have always been trying new things and keeping those that work. With this systemization came increasingly sophisticated and mathematical techniques for teasing out subtle influences and refining scientific models to discover truly universal principles.

The testing of scientific models draws on the same tools and techniques used in the previous steps. The cleanest and simplest example involves the use of a model to predict the results of an entirely new experiment, but we can also test models by repeating the original experiments with greater precision or by extending a model developed in response to one observation to explain the results of another, preexisting observation.

The advance of science is an iterative process, looping repeatedly through the first three steps: looking at the world prompts scientists to think of new models, which are tested by new observations, which prompt new or refined models, which are tested by new observations, and so on. In this section, we'll look at some examples of everyday activities that involve testing our models of the world, and we'll examine some scientific discoveries in which the testing process led scientists to surprising realizations about the universe in which we live.

Chapter 8

QUANTUM CROSSWORDS

When you have eliminated the impossible, whatever remains, however improbable, must be the truth.

—Sherlock Holmes, *The Sign of the Four*

On June 16, 1943, the iconic movie star Charlie Chaplin married Oona O'Neill, daughter of the playwright Eugene O'Neill. Although she was his fourth wife and they were twenty-seven years apart in age—he was fifty-four, and she had just turned eighteen—they remained happily married until his death in 1977 and had eight children together.

I know this not because I have any great interest in early Hollywood movies or Chaplin in particular—though *The Great Dictator* is an excellent movie. I know this because for several years starting in the mid-1990s, I started every morning by doing the *Washington Post* crossword puzzle.

American-style crosswords, like the *Post*'s, consist of a square grid with some squares shaded. The unshaded squares, which tend to form large, contiguous blocks, are to be filled in with letters, which are arranged to form words when read from left to right and from top to bottom. The puzzle is accompanied by a list of numbered clues defining the individual words filling in the grid, both across and down.

Crossword puzzles are wildly popular, and any daily newspaper will include at least one new puzzle every day. Airport bookstores are well stocked with books of puzzles for every level, and there are annual competitions for writing and solving crosswords. The Sunday *New York Times* crossword puzzle has a sort of iconic status as a marker of a certain kind of intellectual.

As you might imagine, finding suitable sets of words that when placed next to one another form still more words is quite a challenge, and good crossword-puzzle writers are a rare breed. Puzzle writers will sometimes resort to strange abbreviations and acronyms to fill out a grid, but they also tend to gravitate toward unusual names, particularly those with large numbers of vowels. Thus, my knowledge of Oona O'Neill, who was a favorite of one of the writers for the *Post*'s crossword puzzle—*O*, *N*, and *A* are among the most common letters in English—and clues of the form "Mrs. Charlie Chaplin" occurred regularly during the years when I was a daily reader of the *Post*. (Another person who regularly turned up in these puzzles and whose existence I would not otherwise have been aware of is the Israeli politician Abba Eban.)

Of course, the first time I ran across one of these clues, I had no idea of Ms. O'Neill's unusual name. And as it was the 1990s, I wasn't able to look it up on *Wikipedia* (which was launched in 2001). Instead, I had to deduce her name using the scientific process: While I had no chance of guessing her name directly, I used the crossing clues to fill in all the letters. And since those answers fit together perfectly, I knew that I had the correct answer for her name. In keeping with Sherlock Holmes's famous dictum quoted at the start of this chapter, once you've filled in all the other clues correctly, whatever remains, however improbable, must be the word you're looking for.

This is a process that millions of crossword aficionados go through every day for fun. The process also captures the essential elements of science. Every time we approach a problem in science, we are trying to discover new information about how the universe works. We do this by recognizing patterns in experimental results that indirectly hint at the information we really want, while we fit those answers into the constraints imposed by other observations and existing scientific theories. This process works to provide us with the best information we have about the universe we live in, even when it sometimes leads us to places that seem improbable in the extreme.

CROSSWORD PUZZLES AND QUANTUM HISTORY

A good starting place for discussing the history of quantum physics is Manchester, England, in 1909, with Ernest Rutherford, of the infamous biologists-are-stamp-collectors jibe. Born on a farm in New Zealand, Rutherford won a scholarship to Cambridge, where he studied under the great J. J. Thomson at the Cavendish Laboratory. He distinguished himself as a great experimental physicist, and by 1909, Rutherford was one of the biggest of the Big Names in physics. He had received his Nobel Prize the previous year for the discovery that alpha particles are the nuclei of helium atoms and that radioactive decay leads to the transmutation of elements. Rutherford was hardly one to rest on his laurels, though, and what might be his greatest discovery was still ahead.

Having determined the identity of alpha particles, Rutherford turned to using them to study the structure of matter. The decay of radium spits out a fast-moving alpha particle, and Rutherford directed a beam of these particles at a thin foil of gold to see what happened when they interacted with gold atoms. By measuring the deflection of the alpha-particle beam,

Rutherford and his students hoped to learn something about the internal structure of atoms, which were known to contain electrons (a discovery made by Thomson in 1897) but were otherwise a mystery.

The state of the art in alpha-particle detection in 1909 was a glass window coated with a small amount of zinc sulfide, which would emit a small flash of light when struck by an alpha particle. These flashes were very faint, and experimental data collection demanded long hours of staring at the zinc sulfide screens through a telescope, which was hard, tedious work. So, in the manner of distinguished scientists from time immemorial, Rutherford delegated the task to his subordinates, specifically his postdoc Hans Geiger and a promising undergraduate named Ernest Marsden.[*]

As part of their experiment, Marsden and Geiger looked for alpha particles hitting their detector when it was on the same side as the source. They expected to see very few particles, as the thin gold foil should not have proved much of a barrier to the high-energy alphas. To their surprise, however, they found large numbers of alpha particles that apparently hit the gold and then bounced backward, even almost directly backward. Rutherford later described this result: "It was quite the most incredible event that has ever happened to me in my life. It was almost as incredible as if you fired a 15-inch shell at a piece of tissue paper and it came back and hit you."[†]

Rutherford, Marsden, and Geiger had no idea of the implications when they started the experiment, but their strange

---

[*] Possibly as a result of this experience, Geiger went on to invent the Geiger counter as an alternative and less annoying means of detecting radioactive particles.

[†] Quoted in E. N. da C. Andrade, *Rutherford and the Nature of the Atom* (Garden City, NY: Doubleday, 1964), 111.

result was the first step toward completely revolutionizing our understanding of how the universe is put together.

So, what is so incredible about this result, and how does it tell us anything about the nature of atoms? At the time, the most popular model of the atom was Thomson's "plum-pudding" model, which involved electrons embedded in an amorphous, positively charged mass, like raisins inside a gelatinous dessert. Rutherford realized at once, however, that Marsden and Geiger's result was completely inconsistent with the plum-pudding model, thanks to the Nobel laureate's knowledge of the physics of collisions.

For an alpha particle to bounce straight backward from a collision with a gold atom, it must be hitting something very heavy and very dense inside the gold atoms.[‡] If the atom were a diffuse positive mass, the alpha particles would punch straight through, hardly deflecting at all, and Marsden and Geiger never would have seen any particles bouncing back to their detector. On hearing Marsden and Geiger's results, Rutherford very quickly realized that the data said something surprising about the structure of the atom. In 1911, he presented the first version of the atom that children nowadays learn about in grade school: a positively charged nucleus at the center, containing the vast majority of the atom's mass, with negatively charged electrons orbiting the nucleus like planets in a miniature solar system.

[‡] You can verify this fairly easily if you have access to a pool table. If you place a billiard ball in the center of the table and hit it with a cue ball, depending on the angle of the collision, you can make the cue ball go in many different angles, but never straight backward. To get the cue ball to bounce backward requires a solid target with much more mass, like several billiard balls stuck together, or a bowling ball placed on the table.

REVISIONS AND REVOLUTIONS

Rutherford's model is very simple and appealing and fits the observations made by his students. It also ran into trouble very quickly: Given what was known of physics in 1911, his model couldn't possibly be right. The classical theory of electromagnetic interactions had been well established at that point, having been worked out by James Clerk Maxwell in the 1850s. According to classical electromagnetism, an electron whipping around an atom in a circular orbit should spray light out in all directions. That light will carry off some energy, causing the electron to slow down, and in very short order, the electron should spiral inward and crash into the nucleus of the atom. As this is manifestly not the case—we are made of atoms, after all, and those atoms are, for the most part, perfectly stable—Rutherford's theory cannot be made to fit with classical physics.

Rutherford's discovery put physics in the position of a crossword enthusiast faced with a promising answer that doesn't seem to fit with the other answers. Just as the solution to a crossword will sometimes require the enthusiast to change the past answers, the solution here was to modify the underlying laws of physics. The groundbreaking step was made by Niels Bohr, a young Danish theoretical physicist working with Rutherford in 1913. Bohr's model of the hydrogen atom was a radical departure from classical physics and one of the launching points for quantum mechanics, one of the two great theories of modern physics (the other being Albert Einstein's theory of relativity).

Bohr proposed that there are certain special orbits—states with specific energies—for electrons going around the nucleus of an atom and that, contrary to all classical expectations, the electrons do not emit any radiation in these orbits. An electron in one such orbit will happily remain there essentially forever. Moreover, any attempt to make an atom from an electron in an orbit that wasn't one of the special allowed states would fail. Electrons inside Bohr's atom absorb or emit light only when they move *between* these orbits, and the frequency of the light absorbed or emitted is determined by the difference between the energies of the two states.[*]

Bohr's model may seem like the exact opposite of the process it's supposed to illustrate, but while the model was a radical leap, it did, in fact, fit with previous observations. In addition to explaining the Marsden and Geiger results, Bohr was also able to perfectly reproduce the characteristic wavelengths of light emitted by hydrogen atoms. The allowed orbits he predicted for hydrogen form a "ladder" of increasing energies, and the visible lines measured in the spectrum of hydrogen match the energy differences for atoms moving from the third, fourth, fifth, and sixth states down to the second.

The idea of associating the frequency of light with an energy also had a precedent. In 1900, Max Planck introduced the first quantum theory in order to explain the characteristic "black-body" spectrum of light emitted by hot objects (the red glow of a hot piece of metal, for example). A hot object emits light over a wide range of frequencies, with the largest amount of light occurring at a wavelength that gets shorter as the object gets hotter (as explained earlier, when you heat a piece of metal, it first glows a dull red, then a brighter orange, then

---

[*] The energy difference determines the frequency of the light absorbed or emitted, and also the wavelength. Wavelength and frequency are complementary descriptions of light, with an increase in the frequency matched by a corresponding decrease in the wavelength. Physicists switch between describing light in terms of wavelength and frequency, depending on which is more convenient for the problem at hand.

yellow, then white, as the light emitted moves to shorter wavelengths). Planck had determined a mathematical function that fit the shape of this characteristic spectrum (i.e., the function predicted the amount of light emitted at a given wavelength for a given temperature), but to explain the origin of this function, he had to resort to a mathematical trick. Planck described his hot object as if it were made up of imaginary oscillators that could each emit light of a particular frequency and as if the amount of energy contained in the light depended on the frequency of the light.

Planck's model is referred to as a *quantum* model after the Latin for "how much," because it predicts that light can only be emitted in discrete amounts. The smallest amount of light that could possibly be emitted at a given frequency was equal to a constant (now called Planck's constant, with a value of 6.6261×10^{-34} kilogram-meters squared per second [kg-m²/s]) multiplied by the frequency. If this minimal amount of light exceeded the amount of energy available from the heat contained in the object, then there would be no light emitted at that frequency. This model allowed Planck to explain why hot objects emit very little high-frequency radiation, which no other theory could satisfactorily explain.*

Planck was never entirely happy with this picture, but in 1905, an unknown patent clerk in Switzerland picked up his idea and ran with it. In the second of the four groundbreaking papers he published that year, Albert Einstein used Planck's quantum hypothesis to explain the photoelectric effect. As the

---

* The most prominent contemporary model predicted that the amount of light emitted should increase exponentially as the frequency increased, so even relatively cold objects should emit huge amounts of high-frequency ultraviolet radiation. This is manifestly not true, and the failure was dubbed the "ultraviolet catastrophe" by physicists, which would be a great name for a band, but is not a happy description for a scientific theory.

name suggests, the photoelectric effect uses light to produce electrons: When you shine light on a piece of metal, electrons come out. The effect was first observed in the 1880s, but by 1905, nobody had been able to explain the energy of the electrons that come out, which increases linearly as the frequency increases. Einstein argued that if you imagine a beam of light as a stream of particles (now called *photons*), each with an energy equal to Planck's constant multiplied by the frequency, the result follows naturally. Each photon can knock at most one electron loose by giving up its energy to that electron. Higher-frequency photons have more energy, thus the electrons they knock loose have more energy than those knocked loose by lower-energy photons. While this idea was widely disparaged when Einstein first proposed it, his predictions were confirmed in every detail by 1916.[†] The photoelectric effect was the only specific result mentioned in the citation when he won the 1921 Nobel Prize in physics.

So, while Bohr's model required a radical revision of classical physics, it fit perfectly with the models of Planck and Einstein and the observed spectrum of hydrogen. Like a good crossword answer, it also provided key information for future discoveries. The mathematical condition that Bohr proposed for determining the allowed electron orbits in hydrogen did not appear to have any theoretical justification, but in 1923, a French Ph.D. student named Louis de Broglie provided one.

[†] Planck himself, writing a recommendation letter, said that Einstein had "missed the target in his speculations" about the photoelectric effect, but Planck urged the hiring committee not to hold it against Einstein, in light of his success with relativity. Einstein's predictions were confirmed grudgingly, as Robert Millikan, the American physicist who made the decisive measurement, had set out to disprove the photon model. Millikan won the 1923 Nobel Prize in part for these measurements, though, which is a nice consolation prize.

De Broglie noticed that the theories of Planck and Einstein treated light, which was usually thought of as a wave, as a particle, and the Frenchman asked what would happen if that process were turned around: What if electrons, which were usually thought of as particles, behaved like waves?

This idea has an elegant sort of symmetry to it, and de Broglie further showed that if you apply it to Bohr's model of hydrogen, you find a natural explanation for the allowed orbits: They are the orbits in which an electron wave completing a single orbit around the nucleus comes exactly back to where it started. The peaks of the wave on the second trip around the nucleus fall in the same places as the peaks in the first and reinforce the wave. For other energies, the peaks of the wave on some subsequent trip around the nucleus fall in the places where the valleys were on the first, and the wave destroys itself. This wave behavior on the part of electrons explains why those particular energies are special and why an electron in those orbits is stable.

Of course, de Broglie's suggestion that electrons behave like waves sounds even crazier than Bohr's original proposal, but again, it not only fit with previous results, but also paved the way for future discoveries. In 1927, two separate experiments, by Clinton Davisson and Lester Germer in the United States, and George Paget Thomson in the United Kingdom, directly demonstrated the wave nature of electrons interacting with solid matter. At around the same time, Werner Heisenberg and Erwin Schrödinger developed the full and correct mathematical theory of quantum mechanics. Their theory incorporates all the experiments and theories—those of Planck, Einstein, Bohr, de Broglie, Davisson and Germer, and Thomson—into a single coherent framework and forms the basis of nearly all of modern physics.

Quantum mechanics replaced the familiar laws of classical physics with a new set of far stranger rules, which seem utterly bizarre to people encountering them for the first time—some physicists, most notably Einstein and Schrödinger, found it so philosophically distasteful that they turned their backs on it completely, despite their pivotal role in creating the theory in the first place.* As strange as it is, though, the theory emerges inevitably from piecing together the results of numerous experiments, in the same way that an unlikely name emerges from a set of crossword answers.

NON-QUANTUM CROSSWORDS

The important lesson to take from the history of quantum physics and crossword puzzles is not just that scientists piece results together from multiple sources of information—after all, every great scientific theory has its origin in a multitude of bits of evidence that all point in the same direction. The real lesson from quantum mechanics is that fitting together all the pieces of evidence will sometimes lead to improbable results, even results that require radical modifications of existing science. The quantum rules that replace Maxwell's equations and Newton's laws seem utterly bizarre by the standards of everyday experience, but they're also inevitable once you put together all the evidence.† Just as crossword answers are best written in pencil, all scientific theories are provisional, even ones as brilliantly successful as Maxwell's equations.

---

* For a more complete explanation of quantum physics, see Chad Orzel, *How to Teach Physics to Your Dog* (New York: Scribner, 2009).
† Maxwell's equations and Newton's laws are not false in the everyday sense—they are, in fact, an excellent approximation of the everyday world, and many scientists and engineers never need to go beyond those. The history described in this chapter proved that they are only an approximation, however, and sometimes, a deeper theory is needed to describe reality.

While quantum physics is far and away the most radical of the great revolutions in science, the history of science is full of examples where previous conventional wisdom was displaced by a completely new approach. The philosopher Thomas Kuhn took this radical revision as one of the defining characteristics of science, declaring that science proceeds through a series of "paradigms," worldviews that endure for long periods of normal science, then are suddenly replaced with incompatible paradigms in brief episodes of "revolutionary science." Lots of scientists are somewhat skeptical of Kuhn's description, but his work certainly represented a revolutionary break in thinking *about* science and how it works.*

The usual example given of a scientific revolution (Kuhn himself wrote a whole book on it) is the shift from geocentric models of the solar system (where the Sun and other planets orbit the Earth) to heliocentric models (where the Earth and other planets go around the Sun). A geocentric model seems intuitively obvious, and such models date back to antiquity.† But as astronomy improved, small errors began to accumulate. Small corrections in the form of "epicycles" were added to the orbits of the planets, but as more and better observations became available, the geocentric theory eventually became too baroque, and in 1543, it was replaced by the heliocentric model put forth by the Polish clergyman and astronomer Nicolaus Copernicus. The model was supported by observations by Tycho Brahe and Galileo Galilei in the early 1600s and then was put on solid theoretical footing by Johannes Kepler and Isaac Newton a few

---

* And, incidentally, introduced the phrase "paradigm shift" into the popular imagination, so you have Kuhn to thank for thousands of tedious business presentations.

† Though there are also ancient versions of heliocentric models, for example, by Aristarchos of Samos around 270 BCE.

decades later. By 1700, geocentrism was comprehensively overthrown and the heliocentric solar system was widely accepted.

Another example, this one from pure math, managed an even longer period of dominance than geocentrism. The *Elements* of Euclid was regarded as *the* definitive work on geometry from its publication in 300 BCE well into the nineteenth century. It lays out the fundamentals of the subjects in a small set of definitions, axioms, and postulates that form the basis for everything else, then uses these to prove numerous mathematical relationships relating to lines, triangles, and other geometric figures. Four of the five postulates are very straightforward, but the fifth, having to do with parallel lines, is far more complex. For centuries, mathematicians struggled to find a way to either reformulate the fifth postulate in a more elegant manner or show that it could be proved as a consequence of the other four.

Eventually, in the early 1800s, Carl Friedrich Gauss, János Bolyai, and Nikolai Ivanovich Lobachevsky realized that the fifth postulate could be replaced, and a new sort of geometry on hyperbolic surfaces was born. Some years later, Bernhard Riemann's discovery of another replacement for the fifth postulate led to another new sort of geometry on elliptical surfaces. The realization that the fifth postulate was not essential produced a great outpouring of new mathematics, which was later found to have practical application. Albert Einstein's theory of general relativity describes gravity in terms of a warping of space-time created by matter.[‡] The theory is naturally

---

[‡] John Wheeler, *Geons, Black Holes, and Quantum Foam: A Life in Physics* (New York: W. W. Norton, 2000), pithily described general relativity this way: "Spacetime tells matter how to move; matter tells spacetime how to curve."

expressed in the mathematics of Riemannian geometry. General relativity revolutionized our understanding of the universe, giving rise to Big Bang cosmology and exotic objects like black holes. It's also one of the most carefully tested theories in the history of science, with countless experimental verifications. As strange as the theory may seem, it's an inevitable consequence of numerous observations, and the necessary math would not be possible without discarding the fifth postulate, though it was cherished for better than two thousand years.

FAILED CROSSWORDS

One of my many pop-culture guilty pleasures is the TV show *Ancient Aliens*, which runs on History (formerly called The History Channel) and related channels, which feature "ancient alien theorists" suggesting that all manner of archeological sites can best be described as the result of extraterrestrial influence. The alien "experts" claim that ancient monuments like Stonehenge and the Pyramids of Giza could only be the work of a more advanced civilization, and they interpret all sorts of ambiguous carvings and paintings as depictions of alien creatures and spaceships. These theorists even use alien technology to explain the magic and miracles described in ancient works of literature. My personal favorite is probably the episode in which they speculate that the extinction of the dinosaurs was a deliberate event, with aliens using advanced weaponry to cause the mass extinction at the end of the Cretaceous to clear the way for humans.*

---

* In the ancient alien model, we're often (but not always) held to be genetically modified offspring of the same aliens who have done all this meddling with our history.

Of course, this is all nonsense, but I find it highly entertaining nonsense. The "ancient alien" models are presented through comically overdramatic narration, and the ancient-alien theorists interviewed in the talking-head segments explain their ideas with great verve (and have spawned several Internet memes in the process). The "evidence" they cite is both ludicrously thin and obviously cherry-picked from a vast amount of data supporting more conventional theories. When it's all put together, the complete package is extremely amusing, at least as long as I can avoid thinking about how some people take this stuff seriously.

In addition to being entertaining fluff, though, *Ancient Aliens* serves as a useful cautionary tale appropriate for this chapter. While the history of quantum physics and other scientific revolutions show the importance of following evidence and fitting pieces together to find improbable results, even when doing so requires modifying previous ideas, *Ancient Aliens* shows the dangers of taking this too far, in either direction.

The most obvious failing of the ancient-alien theorists is that they're too quick to resort to the improbable explanation—they jump right to aliens without doing enough to show that more plausible explanations are, in fact, impossible. To be sure, this is an important rule to keep in mind: Just because some questions have improbable answers does not mean that a new question will also have an improbable answer.

But the ancient-alien theorists also fail in the other direction. While they are too eager to discard all of conventional archeology, they are also too eager to hold on to their pet ideas. The greatest failing of the ancient-alien theory is that it's infinitely malleable. Aliens are an all-purpose explanation for everything odd in history, and there is no level of counterevidence that the ancient-alien theorists will accept as refutation of their theory. Even a lack of evidence can be twisted into evidence for the theory—when crucial pieces can't be found, it's because they were deliberately destroyed by the aliens covering their tracks or suppressed by some shadowy government or scientific establishment. If good science is like a crossword puzzle done in pencil, ancient-alien theory is like writing random words on a blank roll of butcher paper—if an answer doesn't fit at first, well, the writer can always change the wording of the clue or add a few extra letters to *make* it fit.

Proper use of science, then, requires a balance. If the evidence requires it, you have to be willing to discard previous ideas, even ones you hold dear, and even if the replacement ideas seem unlikely at first glance. If you can solve a crossword puzzle without cheating, though, you have all the tools you need to understand even the strangest theories of modern science.

Chapter 9

PRECISION BAKING

There are known knowns; there are things we know that we know. There are known unknowns; that is to say, there are things that we now know we don't know. But there are also unknown unknowns—there are things we do not know we don't know.

—Former US Secretary of Defense Donald Rumsfeld

I n a 2000 profile in *The Industrial Physicist* magazine, Shirley Jackson, the first African American woman to earn a Ph.D. from MIT, and the president of Rensselaer Polytechnic Institute, recalled applying for a summer job in a research laboratory.* Although she was interested in science, she had no prior lab experience when she approached the professor she wanted to work for. On hearing this, he asked her if she could cook. She said yes, and he immediately told her she was hired. "To do what?" she asked.

This turns out not to be as outrageous as it might first seem, because as the professor noted, if you know how to

---

* Jennifer Ouellette, "Shirley Jackson Puts a New Face on Physics," *The Industrial Physicist* 6 (June 2000): 22.

cook, you have the basic skills needed to do lab science.* Even more than the physical ability to measure and mix ingredients, you have the right mental toolkit: Cooking, like experimental science, requires careful control over both ingredients and procedures. You need to mix the right ingredients, in the right order, in the right way, or you risk disaster.

I do most of the cooking at our house, partly because my job makes it easier for me—we live about two miles from campus, but over half an hour from Kate's office—but mostly because I enjoy it. I like food, and I enjoy the process of turning a pile of ingredients into a meal. With one exception: I'm not fond of baking.

If you're not someone who spends a lot of time in the kitchen, this might seem like a distinction without a difference—baking is just cooking things in the oven, right? There's a major difference between the two, though, in terms of the degree of precision required. Baking bread, cakes, or cookies requires far more precision than cooking most meats and vegetables. If you're cooking, and you've got a little more of one ingredient than usual or are missing some component of the dish, you can easily adjust on the fly. If you're baking, leaving out a single ingredient can mean disaster—forgetting a quarter teaspoon of salt can make a batch of pancakes inedibly bland, and using too little baking powder can produce muffins as dense as the pan they're baked in.

Baking is a complicated process, involving the interactions of many elements. The ingredients used, the way they're

---

* Jackson had way more than the basic skills. She went on to a distinguished career in research at Bell Labs and then as a professor at Rutgers, then became the first woman and first African American to chair the Nuclear Regulatory Commission, during the Clinton administration. She has racked up a long list of awards and honors, including election to the National Academy of Engineering.

mixed, the temperature at which they're baked, the type of pan involved, and even the altitude of the kitchen can make dramatic changes in the outcome. Success in baking requires control of all of these to a level that isn't necessary if you're roasting meat or making pasta sauce. In his book on baking, *I'm Just Here for More Food*, TV chef Alton Brown writes about spending years trying and failing to match his grandmother's biscuit recipe, before he realized that he was missing a key element of her technique: the arthritis in her hands. Her reduced range of motion meant that she was barely able to mix the ingredients for the dough. He was overmixing the dough, leading to biscuits with the wrong texture; once he eased up and imitated the way she only patted the dough together, he got much better results.

In another attempt to reproduce a historical recipe, Brown reports that he was foiled by the subtle difference between historical buttermilk (that is, the liquid left behind after churning milk into butter) and the cultured buttermilk found in stores, which is produced by a chemical process. The flavor is similar, but the fat content and acidity are different, and the recipe needs to be adapted to account for this.

It's not that cooking is completely slapdash—you can't do just anything and expect it to work out—but it's far more forgiving. Cooking requires careful attention, but it doesn't demand the obsessive attention to detail that baking does. A good baker needs to control not only the obvious issues, but track down things that don't immediately seem like a problem—the "unknown unknowns," in the immortal classification of Donald Rumsfeld.

In the same way that different methods of food preparation require different levels of attention to detail, different sciences require different levels of control. Most sciences are like cooking, requiring care to control or document the conditions of an experiment or observation, but nothing too extreme.* The analogue of baking, however, is the precision-measurement subfield of physics, where scientists control the experimental parameters at a level that puts even the most obsessive baker to shame.

DETAILS MATTER

Physics is particularly suited to precision measurement, not because it's lacking in complicating factors, but because it's so sensitive to them. The field is concerned with the simplest and most fundamental interactions between objects, and sorting those out from complex, human-scale experiments is a very difficult task. This difficulty traces back to the earliest days of physics, dealing with Newtonian mechanics, the study of the motion of everyday objects.

The starting point of mechanics is Newton's first law of motion, which can be stated as "An object at rest will remain at rest, and an object in motion will continue moving in a straight line at constant speed, unless acted on by a net external force." Many students, even those who dutifully memorize this statement, find it difficult to apply the concept consistently: Presented with an object moving at constant speed, they immediately assume that there must be some force pushing it in the direction of motion.

This misconception is based on everyday intuition: If you look around you, moving objects are almost always subject to

---

* In many sciences—most of biology, geology, and astronomy—it's impossible to directly control the environment, so scientists have to use careful record-keeping to tease out the effect of natural variations in the key conditions.

a pushing force. If you want to move a chair across the room, you need to push it the whole way, or else it will stop short of the goal. When you drive a car down the highway, you need to keep one foot on the gas to maintain your speed. Real objects don't continue moving for long if they're not pushed.

Up into the 1600s, this behavior was thought to reflect natural law. Aristotle held that the natural state of all objects was to be at rest, and any object pushed into motion would quickly return to its natural state. Even today, you sometimes see the confused claim that Newton's laws are an idealization and don't apply in real-world situations.

In fact, Newton's laws are universal and have no trouble with everyday situations.† The critical element is that last clause: "unless acted on by a net external force." In everyday life, objects are always acted on by external forces. As you move a chair across the room, it experiences forces due to the gravitational attraction of the Earth, and friction from the floor. As you drive your car down the highway, it experiences frictional forces within the engine and from the road, and resistance from the air. When these forces are taken into account, Newton's first law works perfectly: The force you need to supply to keep the chair or the car in motion is necessary to cancel out the forces from the rest of the world, so the *net* force on those objects is zero, and they move at constant speed. As with baking, it's not enough to worry about only the ingredients you supply; you must think about how they come together with factors you don't directly control. Careful cataloging of the interactions between the object of interest and

---

† Provided the object is not small enough to require quantum mechanics or moving fast enough to require relativity. These are easy conditions to meet, however, and cover essentially all situations of everyday interest.

the rest of the world is essential to confirm the basic principles of physics.

Once you're aware of the importance of those additional interactions, you can take steps to eliminate or at least greatly reduce their effect. Newton's laws of motion grew out of observations by earlier scientists like Simon Stevin, Isaac Beeckman, and Galileo Galilei, studying phenomena like balls rolling down ramps. Rolling removes much of the effect of friction, isolating the force of gravity and making its effect more obvious. In more modern contexts, we demonstrate Newton's laws in introductory classes using air tracks and air tables that float lightweight objects above a surface, removing the influence of friction. Probably the most extravagant demonstration of basic physics ever was performed by Commander David Scott of Apollo 15, who dropped a hammer and a feather on the surface of the moon, showing that in the absence of air resistance, they fall at the same rate.

Physics has always relied on accounting for and controlling the experimental conditions to test its universal laws and predictions. And while it's relatively simple to remove the perturbations needed to verify Newton's laws of motion at a basic level, more precise tests of smaller forces demand a greater level of control. This need for precision has been the case since very early in the history of physics. Less than a century after Newton, Henry Cavendish performed the first really great precision measurement and set the bar for future experiments.

THE CAVENDISH EXPERIMENT

The experiment that made Cavendish's name was based on another of Newton's laws, in this case, his universal law of gravitation. According to this law, every object with mass in the universe attracts every other object with mass in the universe. This force has a simple and elegant mathematical form, which depends on the two masses and the square of the distance between them (so, doubling the distance between two masses cuts the force to a quarter of its original value). One of Newton's greatest accomplishments was showing that this mathematical form completely explains the orbits of planets in the solar system.

While Newton's laws were enough to explain the general behavior of the planets, setting the scale of the solar system remained a problem. The orbits of the planets were well known relative to each other, but not the absolute sizes—astronomers had measured the ratio of the radius of Mars's orbit to the radius of the Earth's orbit, but not the exact value of either. Numerous methods were employed to try to get the scale, most successfully observations of the transit of Venus, when the planet passes in front of the Sun as seen from Earth. The precision of these measurements was limited, in part, by the gravitational attraction of the mountains that were, then as now, often the site for astronomical observations. This force could slightly deflect the plumb lines that astronomers used to establish the vertical, introducing some error.

The scientific societies of the day, particularly Britain's Royal Society, mounted expeditions to "weigh" mountains and thus quantify their gravitational effect, but the necessary measurements were difficult to make and interpret. In an effort to simplify the process, a new experiment to "weigh" the Earth was launched by John Michell and Henry Cavendish. The idea of the experiment was simple: to measure the gravitational force between two objects of known mass and compare that force to the weight of those same objects due to the gravitational attraction of the Earth. Combined with Newton's law of gravitation, this would give a measurement of the mass of the Earth, which could be converted to a density to allow more accurate determinations of the force from mountains and other landscape features. Michell started the measurement, but died before it could be completed, so the final experiment was carried out by Cavendish and set the standard for precision measurements for years to come.

Henry Cavendish was descended from one of the wealthiest and most powerful families in England, but following in the footsteps of his father, Lord Charles Cavendish, who resigned from Parliament to devote himself full-time to the Royal Society, Henry pursued a career in science rather than the more traditional dabbling in politics. The younger Cavendish was excruciatingly shy and reclusive, but the Michell-Cavendish experiment to weigh the Earth perfectly suited him, as the entire project could be undertaken without his leaving the grounds of his family home outside London.* The apparatus was moved to Cavendish's property after Michell's death in 1793, and after several years of rebuilding and testing, the results of what is now known as *the* Cavendish experiment were presented in 1798.

While the concept of the experiment is simple enough, the execution is incredibly difficult, for the simple reason that gravity is an extremely weak force. This might not seem to be the case—after all, gravity is more than sufficient to prevent me from dunking a basketball—but remember, the everyday

---

* Cavendish was so shy, the only known picture of him is a pencil sketch obtained by subterfuge after he refused to sit for a formal portrait. The artist sat down near the place where Cavendish normally hung his hat and coat, and quickly sketched his face in profile as Cavendish picked up his clothing. The profile sketch was inserted between the far more detailed drawings of hat and coat.

force of gravity is due to the attraction of the entire Earth. The gravitational force on a 70-kilogram person due to a 150-kilogram lead sphere, like those used by Cavendish, at a distance of ten centimeters is roughly one-millionth of the person's weight, far too small to feel.

The scheme devised by Michell and Cavendish used a *torsion pendulum*, a thin rod with two lead spheres at the end, suspended from its center by a fine wire. Left perfectly undisturbed, the barbell will hang still with the wire untwisted, but if two large masses are positioned near the ends, the gravitational attraction between the test masses and the ends of the pendulum will cause the wire to twist. The amount of the twist, and the time required for the bar to twist back and forth, gives a very sensitive measure of the forces involved.

The concept of the experiment is very simple, but Cavendish's report on it in the *Philosophical Transactions* runs to fifty-seven pages, only some of which can be attributed to the more flowery language of eighteenth-century science. Much of the paper's length is given over to describing attempts to assess the reliability of the measurement, what would today be called *uncertainty analysis*. The discussion includes some of the earliest examples of two essential procedures for precision measurement: controlling the environment to eliminate the effect of the "known unknowns" (errors introduced by environmental interactions through well-understood effects), and placing limits on the "unknown unknowns" (perturbations from unexpected sources), by deliberately introducing errors.

In terms of known problems, Cavendish rightly recognized that it required extreme measures to avoid perturbing the apparatus, as a slight breeze would be enough to introduce a spurious twisting of the wire. He even points out that an experimenter just standing near the apparatus will naturally cause air currents that could disturb the pendulum: "If one side of the case is warmer than the other, the air in contact with it will be rarefied, and, in consequence, will ascend, while that on the other side will descend, and produce a current which will draw the arm sensibly aside."*

For this reason, the entire apparatus was placed inside a sealed shed on the Cavendish estate, with an arrangement of pulleys to shift the position of the test masses as needed. Cavendish observed the position of the pendulum from outside the shed, using a telescope to peer through a window. He took similar care to limit the effects of other obvious problems: When he suspected a magnetic effect, he replaced iron rods in the apparatus with nonmagnetic, copper rods and rotated the metal test masses regularly, in case they had been polarized by the Earth's magnetic field. When his early results seemed to change over time, he replaced the wire with a stiffer wire that would be less susceptible to degrading through repeated twisting (Figure 9.1). Like a baker taking care to monitor the oven temperature and account for the altitude, Cavendish made sure his experiment had the best possible environment for success.

In addition to anticipating and trying to prevent problems, Cavendish also pioneered a technique, now a standard part of the precision measurement toolbox, that people often find surprising: When he thought of a problem, he took steps to make it *worse*. This seems counterintuitive—it's certainly not something you would usually do while baking—but introducing a large, known perturbation is a good way to ferret out those

---

* Sir Isaac Newton, (M., Pierre) Bouguer, and Henry Cavendish, *The Laws of Gravitation; Memoirs by Newton, Bouguer and Cavendish, Together with Abstracts of Other Important Memoirs*, Scientific Memoirs series (New York: American Book Company, 1900; Google eBooks, http://tinyurl.com /kjfoe2k).

Figure 9.1. The experimental apparatus for the Cavendish-Michell experiment to weigh the Earth.

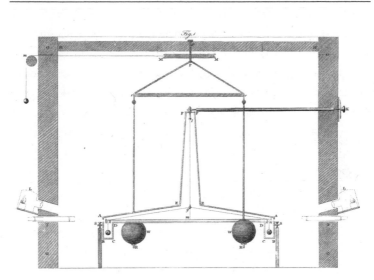

Source: From original Cavendish paper in Sir Isaac Newton, (M., Pierre) Bouguer, and Henry Cavendish, *The Laws of Gravitation; Memoirs by Newton, Bouguer and Cavendish, Together with Abstracts of Other Important Memoirs*, Scientific Memoirs series (New York: American Book Company, 1900); Henry Cavendish, "Experiments to Determine the Density of the Earth," (Phil. Trans. R. Soc. London, January 1, 1798, vol. 88, 469–526).

Rumsfeldian "unknown unknowns." If you're not sure whether small variations in some factor are affecting your measurement, you deliberately introduce a really large change in that factor and observe what effect that has. If it doesn't change your results, you can stop worrying about it, and if it does, you can use the change due to the large perturbation to estimate the effect of the small perturbations you normally face.

Thus, when Cavendish wanted to rule out magnetism as a problem for his measurements, for one test he replaced the lead weights on the pendulum with magnets. On discovering that this did not make a dramatic change in the results, he could conclude with confidence that the much smaller magnetic effects of lead were not a significant source of error.

Making things worse also allowed Cavendish to detect a real source of trouble: placing lamps beneath the weights to heat them turned out to produce a significant change and led him to investigate the effects of temperature more thoroughly, heating the weights and cooling them with ice. He eventually determined that the metal spheres cooled off at a different rate than the wooden case, leading to subtle air currents that affected his results. Having established this as a potential problem and having measured the size of the effect, he took his final set of data to determine the density of the Earth.

All these cautions and cross-checks paid off in a measurement that was difficult to beat and quickly acknowledged as the definitive measurement of the density of the Earth. In the intervening two centuries, the need for such density measurements has become less important, but Cavendish's measurement is still an important part of physics. These days, it's usually recast as the first measurement of Newton's gravitational constant (generally given the symbol G in equations, and referred to as "big G" to distinguish it from g, which is normally used for the constant acceleration of gravity near the surface of the Earth). In those terms, Cavendish's measurement works out to a value of $6.74 \pm 0.04 \times 10^{-11}$ newton-meters squared per kilogram squared ($N\text{-}m^2/kg^2$), within 1 percent of the best modern value. Incredibly, those modern measurements use a variation of the same torsion pendulum technique Cavendish employed. As of 2000, the Eöt-Wash experiment at the University of Washington had used an updated torsion pendulum to measure $G = 6.674215 \pm 0.000092 \times 10^{-11} \ N\text{-}m^2/kg^2$, which is still one of the best measurements of G.

MASS MEASUREMENTS AND THE BOSTON SUBWAY

Each generation of precision measurements raises the bar for the next, and as physics has gotten more complex, the range of effects that need to be considered has expanded. Cavendish had to worry about the thermal effects of a person in the same room as his apparatus, but many modern precision measurements need to worry about disturbances from much greater distances. Every such experiment has its colorful stories about the lengths scientists have to go to for their measurements, but one of my favorite examples comes from Dave Pritchard's group at MIT.

One of the experiments in the Pritchard lab uses an ion trap to make extremely precise measurements of the masses of atoms and molecules. An ion is an atom or a molecule with one or more electrons removed, and with the proper combination of electric and magnetic fields, ions can easily be confined to a small region for extremely long periods—single ions have been trapped for several weeks at a time. Inside the trap, an ion circles around the center at a rate that depends on the mass of the ion and the magnetic field inside the trap. If you measure the frequency at which the ion orbits the trap—something that can be done with exceptional precision—you can determine the mass of the ion with an uncertainty of around a billionth of the total mass.

Of course, this technique requires precise control of the magnetic field inside the trap to extract the mass. Changes in the field from external sources will change the orbit frequency, and while enclosing the system in metal shielding can reduce the effect of small changes in the external field, it's absolutely critical to minimize such changes as much as possible. The MIT group's efforts to understand their system produced a graph that illustrates this very nicely (Figure 9.2).

Figure 9.2. Magnetic field fluctuations in the Pritchard group lab at MIT. The full scale is about 2 percent of the Earth's magnetic field.

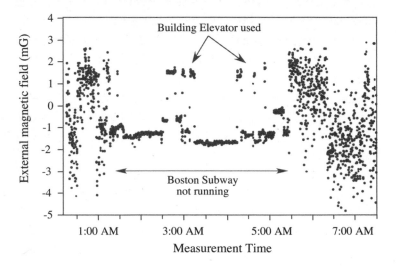

Source: Provided by Simon Rainville. Used with permission.

The graph shows hundreds of measurements of the magnetic field inside the lab, made over about eight hours one night. In an ideal world, these would all give very nearly the same value, as in the stretch around 4 a.m., but as you can see, the actual value fluctuates by a substantial amount over this period, due to sources outside the laboratory. Considerable effort was devoted to finding the source of all these fluctuations, so as to bring them under control and allow ultraprecise mass measurements.

One source of extra magnetic fields was at least somewhat within the team's ability to control: the elevators inside MIT's physics building. These account for the large spikes around 3:00 a.m. and 4:30 a.m. The elevators and their counterweights are large metal objects, and the shifting of their position changes the local magnetic field. The group's detectors were sensitive enough to determine not only when the elevator was moving, but also what floor it was on from the magnetic field measured inside the lab. To get the best data, the team members negotiated with the building custodians to keep the elevators in one place overnight.*

The other big source of changes, though, was well outside the researchers' control and completely unexpected. Graduate students working on the project noticed that the best time for taking data was late at night, roughly 1:30 a.m. to 5:30 a.m. They eventually figured out that the "quiet" hours matched the times when the Boston subway system was shut down for the night. Despite the fact that the nearest station was around a kilometer away, the huge currents needed to move an electric subway train are enough to significantly change the magnetic field in the lab. The system was sensitive enough to detect a single service train moving down the tracks in the middle of the night.

Attaining maximum precision requires averaging many measurements taken over a long period, so the group used to do overnight data runs, scrambling to make as many measurements as possible before train service resumed. Simon Rainville, who worked on the project in the early 2000s, recalls that "Saturday nights were our golden nights for data taking since the subway would start an hour later on Sunday mornings." This, of course, has a certain cost: "It wasn't great for our

---

* Other precision measurement groups face similar problems—the group of Norval Fortson at the University of Washington famously used to disable the physics building's elevator by wedging a trash can in the door to keep it open when the researchers wanted to take data.

social lives, but hey, what wouldn't we do for the advancement of knowledge!"

As annoying as it is to take data at three o'clock in the morning, though, another Pritchard group member, Michael Bradley, described the perturbations from the subway as "a godsend—since once you see them you can compensate for them." The effect of the subways and elevators played the same role that Cavendish's lamps and magnets did, turning "unknown unknowns" into "known knowns." Investigating the source of the perturbations led to a more complete understanding of the stray magnetic fields, greatly improving the team members' confidence in the result. Bradley also recalls sleepless nights worrying that their results were being thrown off by some additional effect that they had not been able to imagine and test; between the two, he preferred the changing magnetic fields.*

Eventually, the Pritchard group worked out an improved technique, using a second ion in the same trap, to correct for the fluctuations induced by trains and elevators. The upgrade let the scientists take data during daylight hours again.† What's more, the improved precision was sufficient to allow a direct test of Einstein's famous $E = mc^2$, showing the tiny change in mass when an unstable nucleus spit out a gamma-ray photon that carried off some energy. The scientists even detected subtle changes in the properties of a carbon monoxide molecule as a result of subtle rearrangements of its electrons.

The case of the Boston subway is just one example of the ways small perturbations affect the most sensitive physics measurements. While the precision-measurement community isn't

---

* As you might imagine, a successful career in precision measurement demands a certain personality type, of which this is a good illustration.

† Rainville: "Our wives were very happy (!) . . . and Dave Pritchard too since we gained over an order of magnitude in precision."

very large, almost every such experiment has its own story of some odd effect that needs to be compensated for. When I was in graduate school, a precision spectroscopy experiment in the lab upstairs included two footprints traced on the floor, because the experiment only worked properly when the postdoc running it stood in exactly that spot.[‡] A modern version of the Cavendish experiment at the University of Washington includes a large stack of lead bricks a few meters to one side of the apparatus, to compensate for the sideways force of gravity due to a large hill behind the physics building. And tiny disturbances even show up in huge experiments—the LEP (Large Electron-Positron Collider) experiment at CERN showed a small periodic change in the energy of colliding particles in their accelerator. The fluctuation was traced to the effect of the tides. This effect is particularly notable because CERN is in land-locked Switzerland, hundreds of miles from the nearest ocean.

Across a wide range of physics, then, we can see the need to account for tiny perturbations to isolate and investigate the fundamental interactions. Like producing high-quality baked goods, precision physics requires exceptional control of the entire environment.

OTHER SCIENCES

The concern with control is not, of course, exclusive to physics. Precision measurements in physics just provide a particularly clear example of a process that is common to all sciences. This is largely a matter of making a virtue of necessity—the phenomena of fundamental physics involve interactions that

---

[‡] He had to stand in that spot to adjust the alignment of the lasers, but his weight bent the floor enough that the beams would go out of alignment if he walked away.

are intrinsically very small and require extreme measures to detect them at all.

In another sense, though, physicists have it relatively easy, as the sorts of interactions they need to consider are relatively limited and universal. As difficult as it can be to track down subtle influences from distant public transit systems, as least the Pritchard group didn't need to account for the mental state of the ions. Experiments in the life sciences confront vastly more complicated systems, and the possible sources of error extend to the psychology of the subjects. When patients complaining of some minor problem are given pills by a doctor, some of the people will show an immediate improvement in their symptoms even when the pills are a harmless placebo with no active ingredients, simply because they expect that the "medicine" ought to make them feel better. This placebo effect can even extend to surgical procedures. In one study, a placebo surgery—in which an incision was made and then sewn shut, with no further treatment—produced the same effects as two popular forms of knee surgery for arthritis. The placebo effect greatly complicates medical studies, making it difficult to determine whether a patient given an experimental treatment improved because the treatment worked or because of the person's expectation that he or she ought to feel better.

The placebo effect is a vexing problem, but doesn't require quite the extreme measures as does precision physics. The solution in life sciences is to compare the experimental treatment not to a group receiving no treatment, but to a group receiving a placebo.* Moreover, to avoid any possible problem

---

* Or better yet, to both a group getting a placebo and a group getting no treatment, though this practice raises some ethical issues for medical researchers.

caused by doctors knowing which treatment is which and sub-
tly cuing their patients, the information about which patients
get which treatment must be hidden from the doctors dispens-
ing treatment as well. Such *placebo-controlled double-blind trials*
are the gold standard in medical research. While this approach
is often presented as a modern development, usually traced
to the work of Claude Bernard in the mid-1800s and William
Rivers in 1908, the history of double-blind trials stretches back
to at least 1835.

In 1834, the head of hospitals for Nürnberg in the kingdom
of Bavaria, Friedrich Wilhelm von Hoven, published (under a
pseudonym) a blistering denunciation of homeopathic medicine. Then as now, homeopathy was a popular alternative
treatment. Upset at being called a quack, the city's leading
homeopathic doctor, Johann Jacob Reuter, issued a challenge,
claiming that the odds were ten to one or better that anyone
taking a properly prepared homeopathic medicine would experience powerful effects.

More than one hundred citizens met in response to this
challenge, and a strikingly modern protocol was developed for
the testing of Reuter's claim. One hundred vials were prepared
and numbered, then randomly split into two lots. One batch
was filled with plain water, the other with a homeopathic di-
lution prepared as instructed by Reuter. A list of which vials
contained which solution was drawn up, then sealed. The vials were passed on to a commission that distributed them to
fifty-four volunteers and prepared another sealed list of which
participants got which vials. Three weeks later, the participants reported what symptoms, if any, they had experienced
after drinking the contents of their vial. At that point, the lists
were unsealed to determine which participants had gotten the
homeopathic solution.

In the end, of the fifty participants who reported results, only eight experienced any notable symptoms, five from the homeopathic treatment group and three from the placebo control. The small total number and even smaller difference between groups led the commission to conclude that Reuter had been wrong and that the pseudonymous van Hoven was correct.* This experiment wasn't quite a perfect double-blind study by modern standards—the reliance on self-selected participants and self-reported symptoms are both problematic—but it clearly establishes the basic principles to be followed in such studies.

More recently, reviews of psychology research have suggested another source of problems, namely, the overreliance on samples drawn from college students in Western countries. One review in 2010 found that two-thirds of studies in leading psychology journals used only American subjects, of whom two-thirds were undergraduates majoring in psychology. The vast majority—96 percent of participants—were from Western countries. This is problematic because this WEIRD sample—Western, educated, industrialized, rich, and democratic—may not, in fact, be representative of humanity as a whole. Some studies have shown that, for example, visual illusions previously thought to be indicative of basic brain functions are, in fact, culturally dependent, and samples drawn from non-WEIRD populations are less susceptible to particular illusions. These findings are recent enough that their full import is still being worked out, but they suggest that yet another unknown unknown has been affecting decades of psychology research.

---

* Sadly, this experiment was not a fatal blow to homeopathy in general, which continues to this day, in spite of failing numerous other, more modern clinical trials in the same manner.

All of these effects demonstrate the crucial importance of exerting a high degree of control over all possible parameters in science. Like Alton Brown, who needed to account for his grandmother's arthritis when he made a biscuit recipe, good scientists need to take everything into account to ensure success.

BEYOND SCIENCE

In 1993, after I graduated from college, I cashed in a bunch of savings bonds and bought a used 1989 Ford Tempo to take with me as I moved to the Washington, D.C., area for graduate school. This car came from a sort of transitional period for Ford and had a number of quirks, among them a tendency to overheat when sitting in traffic during the warmer months.

Over a couple of years, I had this problem looked at several times, at considerable expense. I had the cooling system flushed out, the thermostat replaced, and various other bits and pieces tweaked and adjusted, to no avail. The problem continued to crop up from time to time, most memorably during an August trip to New York City, when I remember sitting in traffic heading into the Lincoln Tunnel with the windows open and the heat running full blast, watching the temperature gauge hover just below the meltdown level.

Finally, the mechanic at the dealership in my hometown where I'd bought the car tracked the problem down, by recreating the full conditions in which the problem occurred: Eugene left the car running in the parking lot for an hour, so the engine began to overheat, and then looked at what wasn't working. The problem turned out to be a fault in the switch that was supposed to turn on a fan to blow air over the hot engine. When I was driving at highway speed, the wind from the car's motion provided ample ventilation to keep the engine cool, and if I ran the heat full blast, the air pulled into the cabin over the engine could keep it from completely boiling over. But when I was driving slowly with the air-conditioning on, the still air would let the temperature shoot up into the danger zone very quickly. The replacement switch cost about a dollar, and I never had that problem again.

The key to the solution is that Eugene, like a good scientist, realized the importance of controlling all the parameters. It wasn't enough to test the thermostat in isolation (that worked just fine) or to jump to a conclusion about the state of the cooling system from my description of the problem. He needed to look at the whole system to see everything in the appropriate context. Deliberately making my car overheat in the parking lot let him find the unknown unknown that was making my summertime drives miserable.

The lesson from all of these stories, from bakers and mechanics, physicists and physicians, and even a widely mocked Bush administration official, is the same: When you're confronted with a problem, you need to make sure you understand and control conditions as completely as possible. Turning unknown unknowns into known knowns is essential, whether you're making physics measurements at the part-per-billion level or just baking a batch of buttermilk biscuits.

Chapter 10

LIKE CHESS WITHOUT THE DICE

SPORTS AND SCIENTIFIC THINKING

Fussball ist wie Schach, nur ohne Würfel.

—Jan Böhmermann spoofing Lukas Podolski

A staple of pop-culture treatments of scientists is the notion of an unbridgeable gulf between nerds and jocks. Like most stereotypes, this portrayal is unfair on both ends. Contrary to popular belief, nerds are not intrinsically opposed to athletic activities—many physicists I know are avid hikers and cyclists, and for years, the international research conference in my field of physics featured a surprisingly intense softball game between labs at MIT and the National Institute of Standards and Technology in Boulder.* This isn't a purely modern development, either—when Ernest Rutherford, who famously disdained theory, hired the theorist Niels Bohr, Rutherford came in for some teasing. The professor sputtered for a moment, then declared, "Bohr's different. He's a *football player!*"†

---

* These games have included at least one Nobel laureate.
† Bohr was an avid football (soccer) player as a young man, and his brother Harald played for the Danish national team.

The "dumb jock" image is nearly as pervasive as that of the unathletic scientist. The former stereotype is largely because professional athletes often come off poorly in media interviews, with some like the endlessly quotable Yogi Berra raising the malapropism to high art. The line often held up as the greatest "dumb jock" quote of all time is often attributed to Polish-born German soccer player Lukas Podolski, describing soccer as "like chess, but without the dice." That's practically a Zen koan, so far past making sense that it begins to seem profound; sadly, it's a spoof by German comedian Jan Böhmermann mocking Podolski. But Böhmermann's ability to fool major newspapers with the fake quote tells you something about the reputation of athletes.

While it's easy to laugh at the verbal stumbles of professional jocks, competitive sports require a surprising amount of mental activity, much of it scientific. In fact, there are few activities more ruthlessly scientific than head-to-head competitive sports like football, soccer, and basketball.* As a player, you must constantly make and update a mental model of all the other players in the game and what they will do next, just as scientists make and update their models of the universe. And every play is an immediate experimental test of your model against objective reality. If your model holds up, you win the game; if your model is wrong, you'll take the blame for the loss.

---

* I use the term *head-to-head* here to mean sports where one player is in direct competition with another, to distinguish them from individual sports where an athlete competes against a clock (track, swimming, etc.) or an established external standard (golf, gymnastics, etc.). This is not to say that competitors in individual sports don't need their brains—far from it—but the mental processes involved are very different.

THE SCIENCE OF COMPETITION

Among major American sports, football is probably the sport whose dumb-jock stereotype is most at odds with reality. While the popular image of football players, particularly those outside the "skill positions" (quarterbacks, running backs, and wide receivers), is of huge guys who are dumb enough to run through walls, the opposite is true. Football players need to remember a remarkable amount of material, and those in the trenches have more to learn than anybody else. It's no accident that offensive linemen have some of the highest scores on the intelligence tests the NFL uses for player evaluation, or that many famous coaches and commentators are drawn from the line and linebacker positions.[†] Some commentators, like ESPN radio host Mike Golic, play off the dumb-jock stereotype for laughs, but very few of the players in those positions are genuinely dumb. Successful interior play in football requires the memorization and application of a huge range of schemes and set plays, with a given player's responsibilities changing dramatically from one play to the next.

Football is the most extreme example of the scientific process in sports, because the constant stopping and restarting allow for the most rigidly set plays of any major sport. Coaches, players, television analysts, and even fans spend countless hours reviewing video of football plays, looking for patterns in the behavior of one team or player, and designing set plays to take advantage of this knowledge. Most other sports don't rely as heavily on set plays as football, but the same process of

---

[†] Coaching legends Mike Ditka (tight end) and John Madden (offensive tackle) played on the offensive line; Bill Parcells and Bill Cowher were linebackers; and Jimmie Johnson was a defensive lineman. Interior players who have made successful careers as commentators include Dan Dierdorf, Tom Jackson, Mark Schlereth, Daryl Johnston, and Mike Golic.

spotting patterns and using them to make strategic decisions goes on even in more fluid sports like soccer, rugby, or my own favorite game, basketball.

I've spent a lot of time over the last thirty years or so playing basketball, mostly in pickup games. As mentioned before, I play in a regular lunchtime pickup game at Union College, where I teach. And while I'm not always consciously thinking like a scientist when I'm on the court, the essential process of the game is scientific.

The most obvious patterns in a game come from playing with the same people many times and learning what they tend to do. A few regular players in our pickup game are left-handed, and I position myself differently when guarding them than I do when guarding a right-handed player. Some other regular players can dribble well with either hand, but will almost always shoot with their right hand; knowing this lets me anticipate that even though they're driving to the basket left-handed, they're going to spin back to their right and get in position to challenge their shot.* Some of these tendencies are even more subtle—one guy I regularly play against doesn't like to back-pedal, for example, so I know that when we run down the court, there will come a moment when he has to look away from me to see where he is. If I make a quick move when he turns his head, I know I can get open for a decent shot, something I've exploited quite a bit.

Strategies to deal with specific players evolve over short time scales, as well. If a guy I'm playing with is shooting and hitting a bunch of jump shots on a given day, I'll adjust to get

---

* To be honest, I'm also prone to this, and often find two or three defenders waiting for me when I turn back to shoot with my right hand.

a little closer to him on defense.† If he's missing long shots, I'll step back a bit more, which lets him shoot but makes it easier for me to stay in front of him if he starts dribbling. If I've turned to my left the last three times I've gotten the ball on offense, I'll fake in that direction, then go to my right, in hopes that the defender will anticipate the same move as in previous possessions and will leave me open for a different shot.

This mental element is a huge part of what makes head-to-head sports so compelling and why I play basketball rather than just going to the fitness center to ride stationary bikes or climb imaginary stairs. I find exercise for its own sake, and even running or biking against a clock, to be boring and repetitive, and I look for excuses to avoid it. The mental aspects of basketball are endlessly fascinating to me, though, and I'll rearrange my schedule to make time to play.

The mental process at work in these sports—testing and refining models of the world—is the essential core of science. It's what enables scientists to slowly converge on reliable information about the world, allowing them to predict what will happen in the future. The process is slower in most sciences than in sports—my mental model of a basketball game is tested against reality dozens of times in a twenty-minute pickup game, while most scientific discoveries require weeks or months of careful analysis. But in at least one area of science, the testing and adjustment happen even more frequently than in basketball: the science of timekeeping.

---

† There's a long-standing argument among statisticians over whether players really do have hot streaks where they hit more shots than you would expect from their overall shooting percentages. It's a tricky question, mathematically, but almost all players, even scientists, believe that if a guy hits his first couple of shots, he's more likely to hit the next, and adjust their play accordingly.

A CLOCK IS A THING THAT TICKS:
A BRIEF HISTORY OF TIMEKEEPING

Stripped down to its most essential nature, a clock is just a thing that ticks: something that performs a regular, repeated action that can be used to mark the passage of time. Any repeating action, from the motion of the planets to the oscillation of light, can and has been used as a clock, and the building of bigger and better clocks is a constant theme in the advance of human civilizations. Scientists advance from one type of clock to another using a standard method for comparing two clocks: You synchronize a new clock with one that is known to be reliable, wait for many "ticks" of the clock, then compare your clock to the reference, and adjust as needed. The operation of a clock thus reproduces the process of science itself, over and over again.

For millennia, the reference "ticking" used to mark time was the motion of the Earth itself: Our planet rotates once about its axis every day and completes one orbit around the Sun every year. Thanks to these motions, we see the Sun rise in the east and set in the west, and over the course of the year, the exact rising and setting positions change in a regular way. Both rotational and orbital motions can be used to track the passage of time.

Many of the oldest human-built structures known today are, fundamentally, clocks. Astronomically aligned sites like Stonehenge use the position of the rising or setting Sun to mark significant dates during a year. These ancient calendar markers are exquisitely precise, even five thousand years later.

While we think of clocks and calendars as separate things, they perform the same function, just on different time scales. In fact, a calendar is a model used to compare the "ticking" due to the Earth's daily rotation with that of its yearly orbit.

When the Sun rises at its northernmost point on the horizon, we know it will be in the same position again approximately 365 days later. We use the daily rising and setting of the Sun to track our progress through the year, and we check the operation of the calendar by testing its prediction of the next day on which the Sun will rise at that northernmost point.

When we check the calendar against the summer and winter solstices, we find that the simplest model—that 365 days pass between solstices—needs a little tweaking. The Earth's orbit around the Sun does not take a round number of days, but a fraction—365.24219 days, give or take a bit. A 365-day calendar will slip by a day every four years or so, and each year the solstice will arrive a little earlier. The scientific model that is the calendar needs regular correction to match reality: thus the system of leap years.

The need for periodic adjustments to keep the calendar in line with the seasons was known as far back as ancient Egypt. Nearly every civilization we know of has had some way of keeping its calendar in synch with the regular arrivals of the solstices and equinoxes, by adding days or even months to the calendar as needed.* The European calendar system in use in most of the world today, with regular leap years containing an extra day, was first instituted by Julius Caesar in 45 BCE, then further refined in the Gregorian reform of 1582. The Gregorian

---

* A few systems don't make formal adjustments to the calendar dates, allowing the months to drift with respect to the solstices. The Islamic calendar is the best modern example, with its months tied to the phases of the moon and occurring at different points in the seasonal year. The month of Ramadan, which devout Muslims mark with a daily fast, drifts by ten or eleven days per year relative to the Gregorian calendar and occurs in the summer, then winter, then summer again on a thirty-three-year cycle.

calendar is wonderfully successful and will keep time with the seasons for thirty-three hundred years before the average date of the solstices slips by one day, but that success requires some complex rules to keep in synch with the seasons.*

Astronomically based calendars track the passage of time on a scale of months and years, but the motion of the Earth can be used to track time on shorter scales as well. The shadow of a vertical object changes direction as the Sun moves across the sky, making a sundial that can be used to track the hours of the day. We can define a day by the time required for the shadow on a sundial to return to the same position, and subdivide the day into hours according to the angle of the shadow. Other types of clocks, such as water clocks and sandglasses, were introduced to allow accurate time measurements at night or in bad weather, and their reliability was established by comparing them with sundials.

The combination of the Earth's rotation and orbit around the Sun introduces some complexity into the operation of a sundial, which becomes apparent once other clocks are available. At higher latitudes, the length of a day changes substantially over a year, and above the Arctic or Antarctic Circles, the Sun can remain above or below the horizon for days or months. Keeping accurate time with a sundial requires quite a bit of scientific sophistication and regular adjustment to keep it in synch with all the motions of the Earth.†

---

* In the Gregorian system, leap years occur every four years, unless that year ends a century, except when the century-ending year is divisible by 400. Thus, 1900 was not a leap year, but 2000 was.

† Generally, sundial timekeeping involves adjusting the position of the hour markers on the sundial to ensure constant-length hours throughout the year. In some timekeeping systems, however, daylight and nighttime hours may be different lengths at different times of year. At one point after the introduction of mechanical clocks, Japanese nobles employed

Sundials and water clocks were the state of the timekeeping art for many centuries, as settled agrarian societies don't demand much more accuracy in timekeeping. As the age of exploration took off and spawned the age of globe-spanning empires, though, accurate timekeeping became critically important. Just as higher-level competition in sports demands more frequent and sophisticated refinements of the game plan, the need for accurate timekeeping spurred the rapid development of new technologies and even a fundamental rethinking of the basic laws of physics.

IF IT'S 3:00, THIS MUST BE SCHENECTADY: TIME AND NAVIGATION

The rise of global empires and accurate clocks go hand in hand because timekeeping is essential to navigation at sea. Thanks to the tilt of the Earth's axis, it's relatively easy to determine one's latitude, by simply measuring the length of a shadow at noon. The longer the shadow, the farther you are from the point where the Sun is directly overhead. This method was used by the ancient Greek philosopher Eratosthenes in around 200 BCE. He accurately estimated the circumference of the Earth by comparing the lengths of shadows in Alexandria and the city of Syene, near modern-day Aswan in Egypt.[‡]

---

specialists to adjust their clocks at sunrise and sunset to keep the same number of day and night hours year-round. In summertime, clocks would be adjusted at sunrise to tick more slowly, giving longer hours during the day, then reset at sunset to tick more quickly, giving shorter hours at night; the adjustments would be reversed in winter.

[‡] There's some dispute about the exact length of the units Eratosthenes used, but even the worst choice of definitions suggests he was within about 17 percent of the modern value, which is pretty impressive for 200 BCE. The most favorable choice is within 2 percent.

Latitude is only one of the values needed to locate your position on the Earth, though. The other coordinate, longitude, is much more difficult to determine.* The most accurate method for determining longitude also uses the motion of the Sun, but rather than measuring the height of the Sun in the sky or the length of a shadow on the ground, longitude measurements require you to know the exact local time, compared with some other point. As you move east or west, the time of sunrise, noon, and sunset changes—this is the reason for the modern system of time zones. If you measure the exact time of local noon (the instant when a shadow cast by a stick is at its shortest) at an unknown location, and compare it to the exact time of local noon at some point whose longitude you know, the difference between those two times tells you the difference in your longitudes. Determining that difference, though, requires accurate timekeeping to determine not just the local time but also the time at some distant location.

The inability to accurately determine the longitude for ships at sea led to some spectacular naval disasters, and in 1714 the British government offered a prize of twenty thousand pounds, a vast sum of money at that time, for a method to determine longitude that would give a ship's position within thirty nautical miles. While all manner of wild schemes were proposed, some based more on magic than science, the most successful approaches fell into two classes. The *method of lunar distances* relied on the fact that the Moon is visible from all over the Earth, and its motion against the background stars follows a predictable pattern. Sailors at sea could measure the position of the Moon

---

* Strictly speaking, to completely specify your position also requires you to know your distance from the center of the Earth, but only a tiny number of humans have ever been far enough off the surface for this distinction to become important.

relative to prominent stars and use that position to determine the time at, say, the Royal Observatory in Greenwich, provided they had sufficiently detailed tables predicting the position of the Moon at particular times for months in the future.

The other method is simpler in concept: Just build a mechanical clock that keeps accurate time over long periods, and use it as a reference. Such a clock was no trivial matter in the 1700s, though, as obvious as it seems to modern audiences. Although one English clockmaker, John Harrison, managed to make a marine chronometer in 1761 that met the criteria for the longitude prize, it required almost superhuman effort. Along the way, Harrison had to invent new technologies to correct for changes in temperature and humidity, and he used bits of diamond to provide low-friction surfaces and avoid the need for lubrication.[†] For a long time, the method of lunar distances was simply more practical—as clock technology improved, shipboard chronometers eventually replaced lunar distance measurements, but the principles of celestial navigation are still taught at the US Naval and Merchant Marine Academies, and the US Naval Observatory continues to produce a nautical almanac with the necessary data about the positions of celestial objects.

Ships at sea aren't the only connection between timekeeping and travel, though. As railroads and telegraphs spread around the world in the 1800s, the need for accurate timekeeping became even more acute. Prior to about 1900, time was a local matter, based on the motion of the Sun, with individual cities each setting the time for their own immediate

---

[†] Harrison's clocks and his long battle with the Board of Longitude are described in Dava Sobel's best-seller *Longitude: The True Story of a Lone Genius Who Solved the Greatest Scientific Problem of His Time* (New York: Walker, 1995), though her treatment favors Harrison in a way that some historians of science regard as unfair.

neighborhood. As it became necessary to coordinate railroad traffic over long distances, however, this patchwork of local hours began to cause problems, including fatal train crashes when confused railroad operators mistakenly sent trains in opposite directions down the same stretch of track.

The nearly (but not quite) instantaneous transmission of telegraph signals down the lines that ran parallel to train tracks made it possible to coordinate time over entire continents. The laying of transoceanic telegraph cables extended this capability to a truly global network, with clocks worldwide synchronized to within a fraction of a second. Establishing a convention to specify precisely what time it is at every position on the Earth's surface required years of careful diplomatic negotiation, but eventually produced the modern system of time zones. Along the way, careful consideration of timekeeping and the process for synchronizing distant clocks led the French scientist and philosopher Henri Poincaré and a then unknown patent clerk in Switzerland by the name of Einstein to reconsider the very notion of time. Their insights eventually led to Einstein's theory of relativity, one of the greatest intellectual achievements in human history.*

TURN LEFT AT 12:19:35.167534237: ATOMIC CLOCKS AND GPS

Modern navigation continues to rely on accurate timekeeping, though computer technology hides most of this process from us. Any time you use a global positioning system (GPS)

---

* The process of clock synchronization and the path from there to relativity are described in detail in Peter Galison, *Einstein's Clocks, Poincaré's Maps: Empires of Time* (New York: W.W. Norton, 2003). For more about the theory of relativity, try Orzel, *How to Teach Relativity to Your Dog.*

receiver—whether it is in your car, in a handheld unit, or on a smartphone—you are using the great advances in timekeeping technology of the twentieth century.

GPS is a network of satellites orbiting the Earth, each satellite containing an atomic clock. The satellites broadcast coded signals that identify the particular satellite and give the time according to that satellite's clock. A GPS receiver on Earth picks up the signals from at least three of these satellites and compares the times reported by the satellites. Since we know from Einstein's theory of relativity that light travels at a fixed speed, knowing the difference in the times received from each of the satellites allows us to determine the distance between the satellite and the receiver. This, in turn, allows the computer in the receiver to work out its position on the surface of the Earth through a process of trilateration.

A much slower analogue of the process goes something like this: If I want to tell some friends how to locate Schenectady, New York, using a map showing only major cities, I can start by telling them that it takes me three hours to drive to New York City. That tells them that I live somewhere on a circle with a radius of around 180 miles, encompassing points in New York, Massachusetts, Rhode Island, New Jersey, and Pennsylvania, which isn't all that helpful by itself. If, however, I add the additional information that it also takes three hours to drive to Boston, that tells them I must be at the point where that initial 180-mile circle around New York City intersects a similar circle centered on Boston. That narrows the choices down to one of two points: Schenectady or somewhere in the middle of the Atlantic off the coast of Long Island. Common sense would then allow them to decide where I actually live, but adding a third circle representing the four-hour drive to Montreal specifies my position on the Earth absolutely: There is only one point where all three of those circles intersect (Figure 10.1).

Figure 10.1. Locating Schenectady, New York, by measuring its distance from New York City, Boston, and Montreal. The trilateration process is similar to the way GPS works.

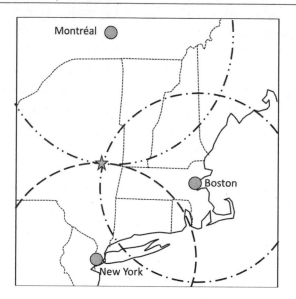

The process for GPS is similar: Knowing the delay between the time signals received from at least three different satellites allows the receiver to calculate the distance to each satellite.* As the orbits of the satellites are well known and carefully maintained, this calculation locates the receiver at the intersection of a set of spheres centered on the various satellites, specifying the position as a single point on the surface of the Earth. In an open area with a clear view of the sky, GPS signals reliably locate a receiver to within a few

---

* Most receivers use at least four signals, treating one of them as a reference clock that is used to determine the distance to the other three, but that's a technical detail.

meters, enabling real-time navigation for hikers and drivers in unfamiliar locations. Tall buildings and other obstacles can complicate this process, but additional information from cellular phone networks can compensate, to the point where a lost-phone-locating app can tell me exactly where in my yard I dropped my phone.

Of course, such a navigation system is only as good as the clocks making it up, and the clocks in GPS satellites need to be very good, indeed. Light travels very nearly one foot per nanosecond (the rare case where American units of measure are more convenient than metric units), so determining your position to within a meter requires clocks that can keep time to within about three nanoseconds over many years in space, since there are no easy opportunities to repair or replace the clocks.

To obtain the necessary timing accuracy and stability, the GPS satellites contain clocks based not on mechanical motion, but on quantum mechanics. Since Niels Bohr's model of hydrogen in 1913 (Chapter 8), we've known that electrons in atoms occupy discrete internal states and that electrons move between these states by absorbing and emitting light. The frequency of the light depends on the energy difference between the initial and final states of the electron, which is determined by the universal laws of physics. Every atom of a given element in the universe is absolutely identical and will absorb and emit the same frequencies of light, making atoms an ideal reference for a clock based on light.

The official definition of the second for the international system of units is the time required for 9,192,631,770 oscillations of the light associated with a particular transition between two states in cesium. All other time and frequency measurements ultimately refer back to this one transition in cesium.

The operating principle of a cesium atomic clock is the same as for any other timekeeping device: The clock is first synchronized with the atoms, then allowed to operate freely for a time, and then measured again, and the frequency is corrected. In the best atomic clocks, several million cesium atoms at a temperature of a few millionths of a degree above absolute zero are prepared in the lower-energy state of the clock transition, then launched upward inside an ultra-high vacuum chamber. As they head up, they pass through a microwave cavity, where they interact with the light, then fly up above the cavity for about a second before falling back down through the cavity, interacting with the light a second time. After this second interaction, the state of the atoms is measured, to see how many of them have moved to the higher-energy state due to their interaction with the light.

This two-interaction technique, developed by Norman Ramsey in the 1950s and for which he shared the Nobel Prize in 1989, uses another unique property of quantum physics to achieve exceptional measurement precision. The first interaction with the light is arranged to put the cesium atoms in a superposition of both of the clock states *at the same time*. This quantum superposition state evolves in time at the frequency of the light, so the first step is analogous to setting a clock and starting it running. If the frequency of the light in the clock perfectly matches the energy difference between the two states, the second interaction with the light will excite 100 percent of the atoms to the upper clock state; if the clock frequency is off, the fraction of atoms making the transition will depend on both the frequency offset and the time between measurements. This dependence on time is what gives the Ramsey method its power—an atomic clock with a one-second time between interactions measures the frequency to within about 1 Hz out of 9,192,631,770 Hz in a single run of the experiment.

Just as a single possession doesn't make for much of a basketball game, though, a single clock cycle is not enough for a real time measurement. To determine the value of the second, scientists repeat the experiment many times and average the results. A few hours' worth of clock operation gives a value of the second accurate to two parts in 10^{16}—that is, 1.000000000000000 ± 0.0000000000000002 seconds. If such a clock operated continuously, it would run for more than 150 million years before gaining or losing one second.

It may seem like atomic clocks have one major advantage over athletes, in that the cesium clock frequency is not a moving target—cesium atoms, unlike opponents in basketball, are not making their own models of the clock operation and aren't adjusting to confound the scientists. While cesium atoms aren't actively malicious, they're not perfectly predictable. As we saw in the previous chapter, numerous small effects (e.g., stray electric or magnetic fields, or additional radiation leaking into the clock) can perturb the energy levels of the atoms. Other factors such as changes in temperature or humidity can affect the performance of the clock's electronics. All these potential perturbations require a constant monitoring of the clock's environment. And on a more fundamental level, the whole idea of describing the performance of an atomic clock in terms of gaining or losing a second is problematic, because it presumes the existence of some single, absolute time that the clock is compared with.

In fact, Einstein's theory of relativity tells us that the passage of time is different for different observers: A clock that is moving will "tick" slower than a clock that is standing still, and a clock that is near a massive object will tick slower than a clock out in space, far away from anything. For most applications, this shift is too small to measure, but it comes into play in GPS navigation: The GPS satellites are in orbit high above the Earth and are moving very fast and thus tick at a different rate than identical clocks on the ground. The scientists and engineers who designed the system understood this and built in a correction factor; without it, the GPS clocks would gain 0.000038 seconds per day relative to clocks on the ground, and position measurements using the system would wander off by up to eleven kilometers per day.

The effects of relativity are too small to introduce significant errors for ordinary earthbound clocks, but scientists are never willing to settle for ordinary, and they continue to design new and better clocks. Atomic clocks in labs at different elevations tick at measurably different rates, requiring some negotiation to combine different clocks to make a truly international time standard.* And physicists at the National Institute of Standards and Technology (NIST) in Boulder have developed experimental clocks that are based on a single ion of aluminum and are so sensitive they can see the effects of relativity. Comparing two aluminum ion clocks clearly shows a difference between the rate for a clock at rest and one where the aluminum atoms are moving back and forth at only a few meters per second (the speed of a brisk walk), or when one clock was raised thirty-three centimeters (a bit more than a foot) above the other.

---

* The International Bureau of Weights and Measures maintains several time standards, combining data from atomic clocks in many countries and incorporating different degrees of correction. The official time established by treaty is Coordinated Universal Time. As a compromise between the English name and the French name (Temps Universel Coordonné), the term is abbreviated UTC, which doesn't make sense in either language.

The science of timekeeping, then, constantly recapitulates the process of science itself. At each cycle of an atomic clock, scientists are effectively making a prediction about their microwave source—a prediction that they then immediately test. If their model is correct, they go on to the next cycle of the clock; if the measurement shows that the frequency was wrong, they make a small correction and try again. This process repeats over and over as long as the clock is running, and produces clocks accurate enough to test the most exotic theories of time, and enables GPS systems to guide even the most stereotypically confused athletes to their destinations.

OTHER SCIENCES

The core process of the science of timekeeping is just a replication of the process of experimental science in general. The distinction between timekeeping and other subfields is just the frequency with which small adjustments are made—an atomic clock in regular operation is checking and refining its prediction of the cesium microwave frequency every second or so, the whole time it's running. Most other experiments operate far more slowly, but perform the same essential function.

Interestingly, this same process of rapid testing and refinement also comes up in a purely theoretical context in biochemistry. In quantum mechanics, it turns out that only a handful of problems can be solved exactly to produce a simple equation you can write down with pencil and paper. Any physical situation more complex than these few simple examples has to be worked out numerically, using a computer to generate an approximate solution.

These calculations can become extremely complex even for fairly simple problems—while the exact energy states of a hydrogen atom can be worked out as a homework problem in undergraduate physics classes, calculating the energy states of the next simplest atom, helium, is a significant challenge. The issue is the interactions between particles: The energy of one of helium's two electrons depends not only on its own position relative to the nucleus, but also on the position of the other electron. But that electron's position depends on its energy, which also depends on both the nucleus and the first electron, making the problem a complex tangle of interactions. And the problem gets exponentially worse as you add more interacting particles, to describe heavier atoms, or molecules in which many atoms are bound together. The situation is especially bad for complex biological molecules like proteins, which start as long chains of amino acid molecules, but fold themselves into complex shapes. As the exact shape of the resulting molecule is critically important for its function in a living cell, a great deal of effort has been directed at this protein-folding problem.

One of the common methods for attacking such problems is through a series of rapidly refined approximations. As a starting point, theoretical chemists and physicists will assume some plausible-seeming set of positions for the components making up a molecule, then calculate the energy and position of the electrons taking those positions as fixed. The distribution of electrons changes where the pieces "want" to be, though, so the next step is to take the electron distributions as fixed, and calculate new positions. The new positions then serve as a starting point for a new round of calculations of electron distributions, which determine new positions, which determine new electron distributions, and so on. The calculation continues, each step refining the model, until the change from one iteration to the next becomes smaller than some desired level of precision, at which point the results can be checked against measurements of real molecules. These calculations can be ferociously complex, requiring hours of supercomputer time to process, and progress has been slow.

One of the most interesting of the many attempted refinements of this process is the FoldIt project, which turned protein folding into a video game. Like the other citizen-science projects discussed earlier in the book, FoldIt uses the superb pattern-recognition abilities of human brains. In FoldIt, volunteers play an Internet game in which they make small adjustments to the configuration of the amino acids in a simulated protein to try to find the folded shape that has the lowest possible energy (which is presumably the shape closest to the real structure). The configurations generated are scored on their energy, and the highest-scoring configurations then serve as the starting point for other players to make their own adjustments. As of 2012, almost a quarter million players had registered with the site and played the various games. The best FoldIt results have provided some insight into the structure of proteins involved with HIV/AIDS and an improved design for the enzyme used as a catalyst in the synthesis of certain chemicals.

This process of repeated small refinements is essential to both science and sports. The protein-folding problem is in some ways an even more apt comparison to the process in competitive athletics than timekeeping, as the protein-folding problem operates on two time scales: a rapid iteration within the computer, checking each step against the previous, and a slower comparison between the predictions generated by those calculations and experimental measurements of real molecules. In the same way, organized athletics includes both rapid testing and refinement within a game—each play is a check on the ever-changing model of the game—and more theoretical adjustments made outside the game, through film study.

Even in a fluid game like basketball and a small-time intercollegiate sports program like Union, film study plays a big role in game preparation. Coaches and players pore over recordings of past contests to look in detail at what worked and what didn't—on a few occasions, our lunchtime pickup game has been able to sneak in a few games in the good gym while the varsity team was off watching video. For a more structured game like football, the use of video is even more important. A friend of my parents is an assistant coach for a state championship high school football team, and he devotes hours to reviewing and editing video clips and sharing them with players via a dazzling array of tablet and smartphone apps for individual study and annotation.

Whether the process happens in the lab or in the gym, this practice of testing, refinement, and retesting is a tremendously effective tool. It applies not only to sports and science, but also to any other system involving the complex interaction of many bodies—driving a car in traffic or managing a diverse group of people at work on some problem. We are constantly making models of the world around us, using them to predict the behavior of others, and testing those predictions against reality. Regular and repeated application of this process lets us produce amazingly sophisticated results, whether those are atomic clocks sensitive enough to test general relativity or game-winning baskets to secure bragging rights in lunchtime hoops.

JUST DO IT: SCIENTIFIC THINKING
WITHOUT THINKING ABOUT THINKING

An obvious objection to the analogy between sports and science is that basketball players aren't consciously going through this thought process. Video study is clearly deliberate and scientific, but the actual game play doesn't seem to involve the same deliberate planning. And that's true—in fact, when I'm aware of planning what to do next in a game, I tend not to do very well. Thinking too much slows you down, giving your opponent extra time to react. The best plays seem to happen unconsciously—you catch the ball, make a move, and only think about what happened after the ball goes through the net. That's why Nike's famous "Just Do It" slogan rings true for so many athletes.

Of course, just because you're unaware of thinking doesn't mean that there's no thought involved. A seemingly unconscious play in basketball is the product of long hours spent in the gym over years of playing. Another player in our lunchtime game once asked how I managed to regularly hit a turnaround jumper, because I seemed to release the ball before I could see the hoop to aim it. The answer is, "Because I've taken that shot ten thousand times"—I've played so much basketball over a long period that some shots are basically reflex actions.

That sort of automatic action is necessary in sports because the action happens very quickly, without time for an athlete to consciously process what's going on. Science generally takes place on a much longer time scale, with numerous opportunities for conscious reflection. If anything, scientists are prone to the opposite problem—spending so long thinking about some pet theory that they come unmoored from reality and convince themselves of conclusions they can't really support. This is not to say, though, that science is without its reflexive actions born of long practice. Most scientists, particularly experimentalists, make decisions on the basis of information that they aren't entirely conscious of, simply because they've internalized the results of years in the lab.

When I was in grad school, I worked with a very complex apparatus that was built by a previous student and involved three or four laser systems and four vacuum pumps. Every morning, I went through a long start-up procedure, turning on a bunch of different bits of equipment in a particular order. One of the first times I went through the start sequence myself, I got halfway through, but got hung up when one particular device wouldn't start.

After several attempts to get it going, I went to find the student who had built the apparatus. The second he opened the door to the lab, he walked directly to one of the vacuum pumps, which had stopped, and switched it back on. "That's your problem," he said. "The cryopump was off. Give it about half an hour to cool down, and it'll work."

"How did you know that?" I asked, since I hadn't noticed it at all. He blinked at me, then said "Couldn't you hear it?" The pump in question made a very characteristic clunking noise, which had been absent until he turned it back on. I was new to the lab, so the sound—or lack thereof—didn't register for me among all the many other noises in the lab, but he had spent enough time working with this apparatus that the absence of one sound immediately registered as something wrong, and he knew exactly how to fix it.

Around five years later, I was writing up my thesis and had handed off the day-to-day operation of the apparatus to some new postdocs. One morning, one of them came into my office saying that they couldn't get the apparatus to start up. I headed over to the lab, and as soon as I opened the door, I walked directly to the cryopump and switched it back on. "How did you know that was off?" the postdoc asked.

I blinked. "Couldn't you hear it?"

STEP FOUR

TELLING

ceiiinosssttuv
—Robert Hooke, 1676, anagram for *"Ut tensio, sic vis,"* announcing the mathematical rule for elastic forces

The secret is comprised in three words—work, finish, publish.
—Michael Faraday, on the secret to success as a scientist

After a scientist has constructed a model to explain some natural phenomenon and tested that model with further observations or experiments, the final, crucial step of the scientific process is to tell others the results. Widespread dissemination of results allows other scientists to check the results for themselves and to use the resulting models as the basis for new discoveries.

Although the sharing of results had been going on informally for millennia, publication was the last step to become part of the formal process of science. The open sharing of results is one feature that divides modern science from alchemy—many alchemists went to great lengths to hide their methods and discoveries from their competitors, and as a result, many of their discoveries are lost to history. In *Absolute Zero and the Conquest of Cold*, Tom Schachtman describes a summer day in 1620, when the alchemist Cornelis Drebbel artificially cooled the Great Hall in Westminster Abbey as a demonstration for King James. Drebbel most likely used a precursor of modern air-conditioning techniques, but as he kept his methods secret, nobody is entirely sure.

Reluctance to share results openly carried over into the early days of modern science, with early scientific luminaries reporting some of their discoveries in the form of Latin anagrams. The encryption allowed them to establish priority for their discoveries, while preventing their rivals from using their results, but it led to a good deal of confusion. Galileo Galilei famously reported two of his astronomical discoveries, the phases of Venus and the rings of Saturn, as Latin anagrams, which were just as famously unscrambled by Johannes Kepler. Unfortunately, Kepler rearranged Galileo's phrases to form different Latin phrases than intended, which the German astronomer interpreted as describing completely different discoveries.

Gradually, however, open dissemination of results became an essential part of institutional science, which helped drive the explosion of scientific knowledge that began in the eighteenth century and continues to this day. In this part of the book, we'll look at some aspects of science that draw on familiar communications skills and some hobbies that use the same tools that scientists do.

Chapter 11

SCIENTIFIC STORYTELLING

Plot is a literary convention. Story is a force of nature.
—Teresa Nielsen Hayden

The flip side of the unfortunate idea that ordinary people are incapable of thinking like scientists is the idea that scientists are incapable of doing things that ordinary people do. Scientists in pop culture are often portrayed as hopelessly inept at communicating to others, either through stammering awkwardness (think Cary Grant working hard not to be charming in *Bringing Up Baby*) or willful ignorance of social convention (think Jim Parsons playing Sheldon Cooper on *The Big Bang Theory*). The message that scientists don't have people skills has been presented over and over in different media over a span of decades.

An alternate version—that scientists don't *need* communication skills—is even disturbingly popular with students planning to study science and engineering. One of the most frustrating experiences I have in teaching introductory physics for scientists and engineers is dealing with students who are shocked to find that they are expected to write lab reports in physics class. Some students claim to have chosen a major in science or engineering specifically as a way of avoiding ever writing papers.

The idea that scientists don't have or need communications skills is a deeply unfortunate misconception. The fourth step of the scientific process is the sharing of results, and this sharing demands communication, at least to other scientists. To the alarm of many young graduate students, public speaking in the form of research seminars and presentations at research conferences is an absolutely essential part of a scientific career. And while research papers in academic journals make heavy use of technical jargon in a way that makes them *seem* incomprehensible to nonscientists, clear writing and explanation are highly prized within the scientific community.

It's no accident that many of the best and most successful scientists are also great communicators. When I was in graduate school, I was lucky enough to work for a Nobel laureate and regularly saw four or five others at conferences.* Though their personal styles of writing and speaking are very different, all of these scientists are outstanding communicators in their own way. Most of what I know about how to talk about physics, I've learned from watching them.

One thing they all have in common is an ability to tell a good story about their research. This is true of all the best communicators in science: When they present results, they don't just throw out facts and figures and trust the audience to sort it all out. Instead, they weave the information into a coherent story about what they did, how they did it, and why.

---

* I worked for William Phillips, who shared the 1997 Nobel Prize in Physics with Steven Chu and Claude Cohen-Tannoudji for the development of laser cooling. I was a student in Bill's group when he won the prize, though it was for work done several years before I joined the group.

The exact form varies from one speaker to another—some start with the theoretical background, and others present an anomalous experimental result first before describing the model that explains it—but the best presentations always have a narrative form.

Great scientists are almost always great storytellers, and their talent carries over to interpersonal communications. Albert Einstein's success with relativity was less as an inventor of the theory than as an expositor of it. Other people had the key ideas long before his 1905 papers, but Einstein's clarity of thought and explanation were key to its widespread acceptance. His tremendous international reputation was due not only to scientific success, but also to personal charisma—he was charming and quotable and carefully managed his public image. And in private, he had no difficulty with interpersonal relationships—quite the contrary, as a collection of letters released in 2006 show he had half a dozen extramarital affairs over the years. Another of the founders of quantum mechanics, Erwin Schrödinger, was famous for both the clarity of his lectures and his unconventional personal life. For many years, he openly carried on an affair with the wife of a colleague, which cost him a position at Oxford.

On a less carnal level, great scientists are often successful as administrators and even politicians, professions that also demand a high degree of interpersonal interaction. One high-profile recent example is Nobel laureate in physics Steven Chu, who served as the secretary of energy during Barack Obama's first term as president, effectively managing the diverse activities of the Department of Energy, interacting with the media, and regularly testifying to Congress. Another recent physics laureate, Carl Wieman, also served in the Obama administration, as the associate director of science in the White House

Office of Science and Technology Policy. But the history of physicists in high places goes back centuries, all the way to Isaac Newton, who served as master of the mint for almost thirty years. Despite Newton's reputed lack of patience for those less talented than himself (basically everybody), this position was obtained through the influence of personal friends. While the post might well have served as a sinecure, Newton was in fact very active during this time, presiding over a massive anti-counterfeiting program and the replacement of a substantial amount of currency after Scotland was absorbed into the United Kingdom in 1707. These efforts required considerable political skill, and Newton was extremely successful.

Even scientists who might seem to justify the popular image of the awkward genius turn out, on closer inspection, to be successful communicators. The most famously awkward physicist of the twentieth century was probably Paul Dirac, a man so taciturn that his colleagues invented a fake unit of measurement in his honor, defined as "the smallest imaginable number of words that someone with the power of speech could utter in company, an average of one word an hour." As socially awkward as he was, though, Dirac wrote a famous and much-loved textbook on quantum physics, widely praised for the quality of its explanations. And to the amazement of many of his colleagues, in 1937 he married Margit Wigner, had two daughters, and remained happily married until his death in 1984.*

This chapter, then, sort of inverts the formula of previous chapters. Rather than talking about how ordinary people

---

* Margit was the sister of Hungarian physicist Eugene Wigner, and Dirac was known to awkwardly introduce her as "Wigner's sister, who is also my wife."

employ the mental skills of professional scientists, I'm going to talk about how scientists succeed by employing the communications skills we tend to associate with nonscientists. In particular, I discuss the importance of stories and storytelling for science and how one of the greatest and most unusual theories of modern physics succeeds precisely because of the way it ties into our fascination with stories.

THE SCIENCE OF STORIES

Explaining the world through stories is, of course, nothing new. Stories and storytelling are as fundamental to human nature as science, and every culture we have records of has its creation story and tales of moral instruction. Although our detailed knowledge of ancient stories extends back only a few thousand years, to the beginnings of written language, human fascination with narrative most likely stretches back much further. Stone carvings and cave paintings dating back forty thousand years mix human and animal figures in intriguing ways, and it's easy to imagine that there are stories behind the images.

Indeed, the tendency to seek and invent narrative is a deeply ingrained part of human nature. We see stories everywhere we look. In a classic psychology experiment, people asked to describe short cartoons of geometric shapes moving about a screen used language that attributed intent to the shapes, as if the objects were conscious actors: "The red triangle chased the blue circle off the screen."[†]

Young children live in a world with little distinction between fact and story. As I started writing this book, my

---

[†] F. Heider and M. Simmel, "An Experimental Study of Apparent Behavior," *American Journal of Psychology* 57 (1944): 243–259.

four-year-old daughter was going through a superhero phase. At various times, she identified herself as Strong Girl, Fast Girl, Brave Girl, Smart Girl, Ninja Girl, and Butterfly Girl, and nearly every day, we heard a new story of how her heroic exploits thwarted the plans of various Bad Guys. Now that she's older, her stories have become more and more involved and are a reliable source of parental entertainment.

This fascination with narrative carries over to explanations of how the world works. A large chunk of mythology consists of attempts to impose narrative on the world, by attributing natural phenomena to capricious or vindictive gods and heroes. These stories generally seem quaint and almost comical, as modern scientific explanations of weather in terms of the motion of air and water in the atmosphere are vastly more effective at predicting the course of major storms. And yet, when a weather disaster does strike, it is virtually (and depressingly) certain that at least one fundamentalist religious leader will attribute it to divine vengeance for something or another.

Modern superstition operates on a smaller scale, as well. Every newspaper in America runs a daily horoscope column, which millions of people read and follow. Otherwise highly educated people will behave as if the motion of distant planets had some significant influence over chance events and interpersonal interactions on Earth. The dogged persistence of even readily debunked ideas like astrology shows the power of the human desire to impose narrative on random events.

THE STORIES OF SCIENCE

Storytelling and even myth making have a place in science, too. In learning about physics, for example, a student can hardly avoid hearing the famous stories of Galileo Galilei's dropping weights off the Leaning Tower of Pisa and Isaac Newton's inventing his theory of gravity when an apple fell on him. Of course, neither of these stories is literally true. There are elements of truth to both—Galileo did careful experiments to demonstrate that light and heavy objects fall at the same rate, and Newton did some of his critical work on gravitation at his family farm, while avoiding a plague outbreak in London. But the colorful and specific anecdotes about the origin of those theories are almost completely fiction.

These stories persist, though, because they are useful. They help fix the key science in the minds of students by embedding the facts within a narrative. A disconnected series of abstract facts and figures is very difficult to remember, but if you can weave those facts into a story, they become easier to remember. The stories of Galileo in the tower and Newton under the apple tree help bring home one of the key early ideas in physics by latching on to the power of stories (in fact, most people remember the stories long after they've forgotten the underlying science).

Essentially all successful scientific theories contain an element of narrative: Event A leads to effect B, which explains observation C. Some sciences even have to resist the temptation to impose too much narrative: Evolutionary biologists have struggled for years against the notion that evolution is inherently progressive, working toward some kind of goal.* And one of the great pitfalls of reporting on medical and psychological research is the mistaken assumption that when two phenomena tend to occur together, one phenomenon must cause

---

* Think of the famous cartoon showing a monkey turning into a stooped caveman and then a modern human, as if all of evolution were a directed process toward us.

the other. "Correlation is not causation" is a mantra among scientists and skeptics, for good reason.

The element of narrative can play a major role in the acceptance of a scientific theory, as well. Theories that seem to lack narrative often become controversial. Quantum mechanics introduces an unavoidable element of randomness to physics, an idea that was so profoundly distasteful to physicists like Einstein and Schrödinger that they never fully accepted the theory, in spite of the pivotal role the two men played in launching it. The inability to predict with certainty the behavior of a specific individual particle in quantum mechanics remains one of the greatest philosophical problems of modern science.

Maybe the best example of the power of narrative in the success of a scientific theory also comes from quantum physics. In the late 1940s, three physicists, Julian Schwinger, Sin-Itiro Tomonaga, and Richard Feynman, independently developed the theory now known as quantum electrodynamics (QED). While all three versions are mathematically equivalent and make exactly the same predictions, Feynman's version has become the dominant form, in part because it harnesses the power of stories.

THE STRANGE THEORY OF LIGHT AND MATTER

Quantum electrodynamics is the theory describing the physics of electrons' interactions with light, or with electric and magnetic fields more generally. A theory of QED seems like it ought to be a relatively simple matter—after all, electrons are the simplest particles we know of, and everything we know about them and their properties comes from looking at their interactions with electric and magnetic fields. Indeed, the first quantum mechanical models dealing with electrons were strikingly successful. Bohr's ad hoc atomic model in 1913 (Chapter 8) worked very nicely for hydrogen, and by the late 1920s, Werner Heisenberg and Schrödinger had worked out theories that successfully described the underlying physics. By 1930, Dirac had found an equation that combined a quantum description of the electron with Einstein's special relativity and used it to predict the existence of the positron, the anti-matter equivalent of an electron. In the early 1930s, it seemed like a complete theory of QED was within reach.

Almost immediately, though, problems began to crop up. An electron creates its own electric field, and a complete theory of QED needs to account for the interaction between the electron and its own field. Attempts to include this *self-energy* in calculations, however, proved disastrous. Even the simplest calculation, for the energy of a lone electron moving through empty space, gave a nonsensical answer: infinity.

For a while, these problems could be swept under the rug: By simply neglecting the self-energy and other problematic terms, theorists produced predictions in reasonable agreement with early experiments. This gave some physicists hope that major problems could be avoided. In the late 1940s, though, a new generation of experiments dragged these issues into the spotlight. The massive scientific projects undertaken during World War II had produced huge improvements in technology, particularly for manipulating radio waves, and as the war ended, physicists returned to their research labs with powerful new measurement devices.

With these new technologies available, two problems quickly turned up for QED. First, the strength of the interaction between an electron and a magnetic field turned out to be slightly higher than predicted by the Dirac equation (an effect called the *anomalous magnetic moment* of the electron).

Second, two states in hydrogen that the Dirac equation predicted should have the same energy turned out to have slightly different energies (an effect known as the *Lamb shift*, after one of its discoverers, Willis Lamb).* These effects are tiny—the increase in the electron's magnetic moment is about 0.1 percent, and the Lamb shift is less than a millionth of the energy of the states in question—but both effects were undeniably real and demanded a theoretical explanation. The obvious source of the problem was those infinite terms that were being neglected, forcing theorists to grapple directly with those issues.

The fundamental problem goes back to the central conundrum of quantum physics, namely, the idea that all objects have both particle and wave nature. Prior to 1900, physics made a clean distinction between particles (found in a single, specific place with well-defined properties) and waves (distributed over some extended region, with less sharply defined characteristics). But as we saw earlier in the book, new discoveries led physicists to conclude that electrons, which had been considered particles, behave like waves in some experiments, while light waves turn out to behave like a stream of particles in some experiments. Reconciling these two sets of properties demands a radical revision of classical physics. In the quantum world, particles like electrons can be spread over some region of space, like waves, while light waves can occasionally be localized to a small region, like particles. On top of that, as Dirac showed, physical particles have antimatter equivalents, so material objects can be converted into energy, and vice versa.

---

* Why Lamb gets his name attached to this shift (which he measured with a graduate student, Robert Retherford) but the original discoverers of the anomalous magnetic moment (John Nafe and Edward Nelson) are nearly completely forgotten is a topic for a master's thesis in the sociology of science.

An electron and a positron that come into contact will annihilate each other, turning into two photons of light, or a photon with enough energy can disappear to create an electron and a positron.

In physics terminology, both electrons and light can be thought of in terms of quantum fields, mathematical objects unlike anything in classical physics. Electrons and photons are neither purely waves nor purely particles, but a third class of object sharing some properties of each. One of the consequences of this dual wave-particle nature is that empty space is no longer perfectly empty, but necessarily contains a bit of energy associated with both the light field and the electron field. This *vacuum energy* must be taken into account when you are calculating the properties of the electron. The vacuum energy can occasionally manifest as particles created from empty space, which changes the nature of the interactions between an electron and an electromagnetic field. There are an infinite number of ways these virtual particles can turn up, and attempting to account for them is what leads to the infinite results.

A number of radical solutions were proposed for this problem—redefining or discarding well-established physical concepts or adding new forces to physics—but in the end, the solution turns out to be surprisingly conservative. The key step is recognizing that because we can never observe an electron *without* these virtual particles, the properties physicists measure for an isolated electron aren't its "real" charge and mass, but are what's left after all these interactions are taken into account. The interaction energies that experimental physicists measure are the difference between the energy of the electron with its associated virtual particles and the energy of the electron with its associated virtual particles plus the interaction. The process of *renormalization* fixes the problem by subtracting one infinite energy from another to get a finite value for the measured energy.

This idea was suggested a few times in the early 1930s, but subtracting infinity from infinity is a dodgy business, mathematically. Extremely stringent conditions must be met for the difference between two infinite numbers to give a meaningful result, and it was not clear that QED calculations had the appropriate properties. The standard mathematical techniques then in use weren't up to the job of proving that renormalization was a legitimate path, so early attempts were abandoned. New techniques needed to be brought to bear to solve the problem of infinite results in quantum field theory. Schwinger, Tomonaga, and Feynman were the first to develop and report the necessary techniques, though in very different ways consistent with their own circumstances and personalities.*

Schwinger was the first to succeed, in what seemed the inevitable next step of an already distinguished career. A mathematical prodigy from a prosperous family in Manhattan, Schwinger wrote his first paper on QED at sixteen (though he didn't publish it) and received his Ph.D. from Columbia at twenty-one. The noted physicist Hans Bethe, after meeting the seventeen-year-old Schwinger, wrote to a colleague that "Schwinger already understands 90% of physics; the remaining 10% should only take a few days."†

After learning about the Lamb shift and anomalous magnetic moment at a workshop in Shelter Island, New York, in

---

* The Swiss physicist Ernst Stueckelberg worked out something very much like the final form of QED in the mid-1930s, but he published it in an obscure journal using a confusing system of notation of his own devising, so very few people knew anything about it. Which serves as another reminder of the importance of communication to science.

† Silvan Schweber, *QED and the Men Who Made It: Dyson, Feynman, Schwinger, and Tomonaga* (Princeton, NJ: Princeton University Press, 1994).

1947, Schwinger was captivated by the magnetic moment problem and set about reformulating quantum electrodynamics. In less than a year, he had the answer, and at the follow-up conference in Pocono Manor, Pennsylvania, he presented his work to a select group of the world's leading physicists in a talk that took the better part of a day. Schwinger's presentation was, by all accounts, a dazzling display of mathematical physics. He treated the problem of QED entirely in terms of fields and brought to bear an array of powerful mathematical techniques, using the formal mathematical properties of the quantum fields to classify the infinities involved and show that they could cancel each other out. Schwinger's approach began with very conventional physics, and the mathematical techniques he used, while extremely abstract and formal, were rigorously correct in every detail.

At about the same time, working in isolation in the devastation of post–World War II Japan, Tomonaga developed an approach very similar to Schwinger's. After learning of the Lamb shift from a newspaper article, Tomonaga and his students quickly calculated the correct value and dashed off a paper to Robert Oppenheimer at the University of California in Berkeley, who made sure that it was published and Tomonaga was recognized for his achievement.[‡]

Back at Pocono Manor, the task of following Schwinger's mathematical tour de force fell to another young physicist who had his own solution to the problems of QED. Like Schwinger, Richard Feynman was born in New York City in 1918, but other than that coincidence and exceptional mathematical ability, the two men could hardly have been more different.

[‡] Oppenheimer, of course, led the Manhattan Project to develop the atomic bombs dropped on Japan in 1945, so it's an interesting historical footnote that he was the one to secure Tomonaga's place in the history of physics.

Schwinger was very reserved and formal, always impecca-bly dressed. Feynman, in contrast, cherished (and cultivated) a reputation as a maverick: He played practical jokes, flouted convention whenever possible, and presented himself as "a man without an internal censor, . . . who said exactly what was on his mind the instant it occurred."[*]

Feynman's approach to QED, like Schwinger's, can be seen as an extension of his personality. Where Schwinger started with relatively conventional physics and proceeded with ab-solute rigor and precision, Feynman began with a different mathematical formulation of quantum physics, one that he had developed as a Ph.D. student, and proceeded via unorth-odox methods of his own invention. As a means of keeping track of his calculations, he invented a sort of graphical short-hand, drawing little pictures to stand for complex calculations, and employed tricks whose only justification was that they gave the right answer in the end.

Feynman's presentation at the Pocono Manor conference was an unmitigated disaster. He made the tactical error of try-ing to dispense with formality and to demonstrate his method by doing lots of examples. His techniques were too unfamil-iar, though, and his cavalier approach to mathematics annoyed many in the audience. His graphical shorthand confused most of the listeners and actively offended Niels Bohr, who misun-derstood the pictures as directly contradicting the Copenha-gen interpretation of quantum mechanics that Bohr had spent years assembling and promoting. Perhaps the best indicator of Feynman's failure is an anecdote related by Frank Close in *The Infinity Puzzle*. Some weeks after the conference, as Enrico

---

[*] Robert Crease and Charles C. Mann, *The Second Creation* (New York: MacMillan, 1986). This book is an excellent history of the twentieth-century revolutions in physics.

Fermi's students worked through the notes Fermi had taken on Schwinger's talk, one of them asked what Feynman had talked about. Neither Fermi nor some other colleagues who had been at the meeting could recall anything other than that Feynman had used idiosyncratic notation.

A PICTURE WORTH A THOUSAND EQUATIONS

There's no small irony, then, in the fact that it's almost impossible to find a detailed description of Schwinger's original methods these days.[†] Schwinger's early success won the admiration of all the physicists at the Pocono Manor conference, but in the longer term, Feynman's approach has been universally adopted by physicists, and his graphical shorthand has become one of the iconic images of modern physics.

All three approaches to QED are mathematically equivalent, which was demonstrated by Freeman Dyson in late 1948 after he had spent extended periods working with both Schwinger and Feynman.[‡] Even before Dyson's work, though, it was clear that both Schwinger and Feynman were onto something—Feynman recalled his conversations with Schwinger at the Pocono Manor conference as one of the few bright spots of the experience. Neither man completely understood the other's method, but they both got the same answer for every problem and came away with mutual respect.

If the two theories are equivalent, though, and Schwinger's was more impressive on first presentation, how has Feynman's version of QED assumed such a dominant position that

---

[†] Schweber's *QED and the Men Who Made It* has a comprehensive description of Schwinger's initial method and several subsequent refinements.

[‡] Some physicists feel Dyson's contribution to QED would have merited inclusion in the QED Nobel Prize were it not for the arbitrary rule that the Nobel can be shared by no more than three individuals.

even modern discussions of Schwinger's approach tend to fall back on Feynman's graphical methods for explanation? In the end, Feynman's approach is simply easier to work with— "more user-friendly," in the words of a theorist colleague—in no small part because it taps into our sense of story.

The central problem of QED is that the nature of quantum fields means that there are an infinite number of ways for an electron to interact with light. Extracting a finite answer from a QED calculation requires finding a way to sort and classify these interactions so as to extract only the most relevant information. Schwinger accomplished this using the formal mathematical properties of quantum fields, while Feynman classified the interactions by the stories they tell.

The centerpiece of Feynman's version of QED is his graphical shorthand, now known as *Feynman diagrams*. Each of these pictures tells a story about something that might happen to an electron. The simplest Feynman diagram for an electron interacting with an electric or magnetic field looks like Figure 11.1.

Figure 11.1. Basic Feynman diagram for an electron interacting with an electric or magnetic field.

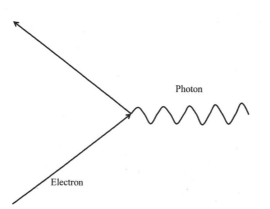

Photon

Electron

The straight lines represent an electron moving through some region of space, and the squiggly line represents a photon from the electromagnetic field. In this diagram, time flows from the bottom of the picture to the top of the picture, while the horizontal direction indicates motion through space, so the diagram is telling a story: "Once upon a time, there was an electron, which interacted with a photon, and changed its direction of motion."

This rather boring story might seem like the complete picture of what happens when an electron interacts with light, but there are myriad other possibilities, thanks to vacuum energy and virtual particles. Even processes that seem to make no sense need to be considered. For example, we can have Figure 11.2.

Figure 11.2. Feynman diagram representing the "self-energy" of an electron interacting with its own electromagnetic field.

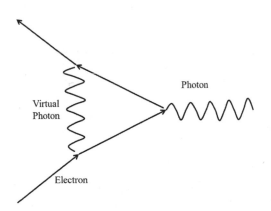

This tells a much more eventful story. Our electron is moving along through space, and just before it interacts with the photon from the electromagnetic field, it emits a virtual photon and changes direction. Then it interacts with the real photon from the field, changes direction again, then reabsorbs the virtual photon it emitted earlier. This represents the problematic self-energy, the interaction between the electron and its own field.

A third possibility has the incoming photon spontaneously create an electron-positron pair.* The pair then mutually annihilates, turning back into a photon, which interacts with the electron. This diagram represents the *vacuum polarization* resulting from virtual electrons and positrons appearing out of nothing (Figure 11.3).

Figure 11.3. Feynman diagram for the "vacuum polarization" resulting from virtual electron-positron pairs.

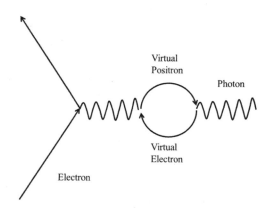

Virtual
Positron

Photon

Virtual
Electron

Electron

---

* A positron can be mathematically described as an electron moving backward in time, another trick invented by Feynman, hence the downward arrow.

Each of these diagrams, and infinite other combinations and permutations, stands for a mathematical calculation. The key to Feynman's method is that there are relatively simple rules for evaluating the contribution each diagram makes to whatever you want to calculate. In particular, for every point in a given diagram where three lines come together, the size of the contribution that diagram makes to the final result decreases by a factor of just over 137.[†]

The diagrams and the simple rules for estimating their size impose a strict hierarchical order on the chaos of infinite possibilities. When calculating quantities in QED using Feynman's method, physicists first ask what precision they require, and then only write down those diagrams that will contribute at the appropriate level. This process removes the need to do an infinite number of calculations (which is impossible by definition) and allows the calculation of finite results at whatever level of precision is needed to match experimental measurements.

These calculations may be very complicated, but the power of the theory is remarkable. The most recent measurement of the anomalous magnetic moment of the electron, one of the quantities that kicked off modern QED in 1947, gives a value of:

$$g = 2.00231930436146 \pm 0.00000000000056$$

The corresponding QED calculation involves nearly a thousand Feynman diagrams, including processes like the one

---

[†] The best current value for this factor, which is represented by the Greek letter alpha and is called the *fine structure constant* for historical reasons, is $1/\alpha = 137.035999084 \pm 0.000000051$. The fact that this is close to but not exactly 137 (itself a somewhat unusual number) drew a few eminent physicists of the mid-twentieth century into wild speculations that were essentially numerology.

Figure 11.4. Feynman diagram for an extremely unlikely process involving four virtual photons and a virtual electron-positron pair.

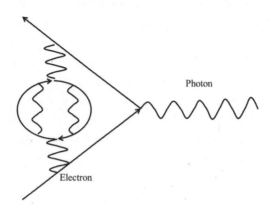

shown in Figure 11.4, with four virtual photons and a virtual electron-positron pair, and agrees perfectly with the experiment. These measurements make QED arguably the most precisely tested theory in the history of science.

The formal mathematics used by Schwinger and Tomonaga were unfamiliar to physicists of 1947, but the mathematical rigor was undeniable and provided a comfortable anchor at Pocono Manor. While powerful, though, the mathematics proved difficult to use—some physicists complained that Schwinger was the only one who could get results by Schwinger's methods—and hard to extract physical insight from. Feynman's more swashbuckling approach to the problem was initially off-putting, but by imposing narrative on the process, it provides a clearer, more intuitive sense of what's going on. Casting the problem in terms of stories about particles interacting in well-defined ways makes it easier to understand the underlying physics and what calculations need to be done. Once Dyson showed that all three methods give the same results, physicists the world over, particularly those of the younger generation, latched on to Feynman's method, and by the 1960s, it was the dominant calculational method for QED.

Theoretical physics has become more mathematically sophisticated since the 1940s, and the more formal mathematical methods used by Schwinger and Tomonaga have been influential in that development. The renormalization procedures developed by all three of the founders of QED have been formalized and extended to new interactions and remain central to theoretical particle physics. Feynman's stories in pictures nevertheless remain essential for an intuitive understanding of the physics of particle interactions, to the point where even discussions of the pre-1947 state of QED rely on Feynman diagrams to explain the problems. The success of QED stands as a testament to the power of stories to make even the most bizarre theories more comprehensible.[*]

---

[*] The telling of stories more or less defines Feynman's career. In addition to his work with QED and groundbreaking research on what are now called *quarks*, Feynman enjoyed a stellar reputation as an expositor of physics. His lectures on QED and other subjects have formed the basis of several best-selling science books, and the *Feynman Lectures on Physics*, created when he taught Caltech's freshman course in 1965, are an enduring classic, still used by educators today (Richard P. Feynman, Robert B. Leighton, and Matthew Sands, *The Feynman Lectures on Physics*, 3 vols. [Reading, MA: Addison-Wesley, 1963–1965]). He also has a larger-than-life reputation as a colorful character, promoted in part through his best-selling collections of autobiographical anecdotes: Richard P. Feynman, as told to Ralph Leighton, *Surely You're Joking, Mr. Feynman! Adventures of a Curious Character*, ed. Edward Hutchings (New York: W.W. Norton, 1985) and Richard P. Feynman, as told to Ralph Leighton, *What Do You Care What Other People Think? Further Adventures of a Curious Character* (New York: Norton, 1988).

CAUTIONARY NOTES

The power of stories to put across complex ideas is well known both inside and outside science. A great deal of scientific research has been devoted to the question of how we interact with stories, and both scientists and nonscientists find narrative within the world and impose narrative upon it. In his 2013 book *The Storytelling Animal*, Jonathan Gottschall summarizes much of this research in support of an argument that storytelling is *the* defining human characteristic. This may seem to conflict with my own argument in favor of science as the essential human activity, but then the line between science and storytelling is not that sharp. The telling of stories—here's a thing that happened, here's our theory why it went that way, here's what we did to test that, and here are the results—is an essential part of the process of science. And the facts and models generated by the scientific process serve as inputs helping shape the stories we tell each other about our world and how it works.

The power of narrative to convey complex ideas, sometimes without the audience realizing it, is well known on a practical level to many people outside science. Unfortunately, this power is employed by some unsavory characters. It's no accident that political speeches tend to be full of anecdotes, often to the exclusion of concrete policy arguments. Personal anecdotes, though they are often misleading, connect policies to people in a way that figures do not, to the endless frustration of social scientists. Marketers, too, are very adept at using stories to sell things, most insidiously to children. Nearly all of the TV shows my children have become interested in—*Jake and the Never Land Pirates*, *Pokémon*, even *Sesame Street*—largely serve as a vehicle for selling an astonishing array of tie-in merchandise. These programs hawk their wares by hooking the kids through engaging stories about colorful characters, who just happen to be available as plush toys or faces on T-shirts.*

Selling through stories is an unfortunate part of science as well. Sometimes this is relatively innocuous, as with the perpetual tension between scientists and the science writers, who many scientists claim oversimplify and hype results to make a better news story. Good stories can be used to sell or even create dodgy science, though. The book *Abominable Science! Origins of the Yeti, Nessie, and Other Famous Cryptids* by Daniel Loxton and Donald Prothero traces the history of some of the world's most famous legendary monsters and shows how the allure of a good story—dinosaurs surviving into the present day! ancient protohumans skulking around the wilderness!—has led many well-intentioned researchers astray. In the case of the Loch Ness Monster, Loxton and Prothero even make a plausible argument that the image of "Nessie" directly originates in fiction, specifically the dinosaur sequences in the 1933 movie *King Kong*.

Distortion of science to fit a story can also have a darker edge, particularly in the social and anthropological sciences, where research findings often have direct political implications. In such contexts, there's a pronounced tendency for researchers to discover results that just happen to fit well with their own ideological preferences and support their preferred policy measures. Highly charged issues like gun control spawn collections of diametrically opposed studies funded by think tanks at opposite ends of the political spectrum. To observers in the middle, both sides seem to be looking for facts to fit

---

* This is hardly a recent development, as the vast collection of Star Wars paraphernalia—figures, toy spaceships, etc.—left in my parents' attic from my own childhood can testify.

their preferred story, rather than collecting evidence first and constructing a narrative afterward.

Politically motivated reversal of the proper order can also operate on a grand scale. In the centuries since the European colonization of the Americas, a combination of ignorance, political expediency, and outright racism has established a particular story about colonization: that prior to the arrival of Europeans, the Americas were only sparsely populated, and the inhabitants did not have a large-scale or sophisticated civilization. This is a very convenient narrative if you happen to be descended from those European colonists, and so it became rather firmly entrenched, with the occasional anomalous element shrugged off or reinterpreted to fit the well-known story.

In the last few decades, though, archeologists, anthropologists, and historians looking closely at the evidence have shown that this convenient narrative is largely false. The dramatic initial conquests—a few hundred Europeans taking vast swaths of territory—had less to do with technological superiority than canny exploitation of local politics and just plain luck.* The sparse and apparently unsophisticated populations encountered by later waves of colonists were the last remnants of populous and sophisticated civilizations decimated by diseases brought by the initial wave—some estimates suggest that more than 90 percent of the population of the Americas in 1491 was wiped out by disease in the space of a few decades.† European settlers weren't moving into lands that had

---

* For example, the Spanish conquest of the Aztec empire was led by Hernán Cortés and famously involved only a few thousand Spaniards with horses and guns, but he won the decisive battles, thanks to an alliance with the Tlaxcalteca, local rivals of the Aztec, who mustered tens of thousands of warriors.

† Charles C. Mann, *1491: New Revelations of the Americas Before Columbus* (New York: Knopf, 2005).

never been inhabited; they were taking over a postapocalyptic wasteland. Archeological evidence shows that, contrary to the traditional story, the original population was quite large and had been managing and engineering the environment on a grand scale prior to the collapse.

The traditional story is slowly being displaced through the construction of a counternarrative, stitched together from bits of research to make a coherent story of the Americas before the arrival of Europeans. The alternative story is as compelling in its own way, though considerably less flattering to European vanity. The bestselling book *1491: New Revelations of the Americas Before Columbus*, by Charles C. Mann, is an excellent example of this use of counternarrative, pulling together a wide range of research and story to paint a more accurate picture.

So, it's important to temper any discussion of storytelling with a cautionary note: The same factors that make stories a powerful tool for connecting with people also make them an exceptionally effective means of deception. It's essential, in science and beyond, to maintain a skeptical position toward stories told by others, but it's even more important to be cautious about the stories you tell. In many ways, the easiest person to fool is yourself, and it can be awfully tempting to adjust a few facts here and there to fit a story you'd really like to be true rather than constructing a story from the facts that you have.

Fortunately, just as telling stories about ourselves is a universal part of social interaction, so too is spotting exaggerations in the stories of others. We all know somebody who believes, despite all evidence to the contrary, that they are a misunderstood genius. An exaggerated version of this type is Bertie Wooster, the narrator of P. G. Wodehouse's classic comic novels. Much of the fun of those books lies in the contrast between Bertie's heroic description of his own role in events and the bumbling reality that comes through between the lines. Very few readers would ever take Bertie's version literally, because we're very good at spotting the gaps between the story he's telling and the story Wodehouse is telling.

In much the same way that the signals of Bertie's buffoonery are clear, the signs of research where the facts are bent to fit a preferred story are not that hard to spot: The results are a little too clearly in line with the political views of the people who paid for the research, and plausible alternate explanations are either ignored or brushed aside too quickly. In science as in anything else, if something sounds too good to be true, it probably isn't. It's also important to apply this same filter to the stories you tell yourself, to make sure your story about what happened is recognizably the same as the stories others would tell with the same information. Regardless of who's telling the story, though, you should always approach stories about science with the same skepticism you would use for stories told by a Wooster.

STORYTELLING BEYOND SCIENCE

One of the best classes I took as an undergraduate was a history course about Vietnam, taught by a particularly good professor. One week, the reading assignment was two translated articles from a Chinese historical journal in the late 1960s. These discussed two unrelated incidents from dynasties hundreds of years earlier and seemed so incongruous I wondered whether the library staff had given me the wrong readings for the course packet.

Talking with my fellow students before class revealed that we had all read the same puzzling articles. The professor started class by asking, "Do you know why I made you read these?" We all said no. "Would it help," he asked, "if I told you that the author of this article was promoted to a leadership post a few months later, while the author of the other was stripped of his positions and died in a forced labor camp?" It didn't help, but certainly got our attention.

The point of the readings, he explained, was that these articles were written during the Cultural Revolution in China, when Mao Zedong was running the country with more or less absolute power and having his political rivals imprisoned or killed. In those circumstances, it wasn't safe to openly advocate particular government actions, because if Mao disagreed, it could mean exile or death. Policy debates shifted underground, then, into venues like the historical articles we had read. One of the two told a story about an ancient emperor who on seeing two rival powers warring with each other left them to fight among themselves and presided over a long period of peace and prosperity. The other told a story about a different emperor who took the opportunity of a conflict between two neighboring states to absorb both, greatly expanding the power and prestige of China. The former was a disguised argument for China to remain neutral in the ongoing war between the United States and the Soviet-backed North Vietnam and to concentrate on building up its internal strength, while the latter was advocating for intervention in Vietnam as a way for China to gain power over both the United States and the Soviet Union. Mao opposed intervention and—unfortunately for the latter author—saw through the veiled opinion and had the author killed.

I mention this as an example of the importance of stories and storytelling to education in general. I no longer remember the names of the specific writers or the emperors in their articles, but the vivid story around the lesson fixed the general point indelibly in my memory. Had we merely been assigned a textbook reading or an article laying out Mao's position on intervention in Vietnam, I probably would have forgotten it entirely, but more than twenty years later, I still remember the moment when everything clicked into place and the whole thing made sense.

Physics doesn't provide many opportunities to be quite that dramatic, but I think about that class a good deal when I'm teaching. It's a great example of the effectiveness of building a key point into a larger story and how narrative can help convey ideas more effectively than a simple recitation of facts. This general principle is a big part of recent thinking about effective teaching in physics, which shows that students often learn better through indirect methods where they discover *how* we know key physics concepts rather than just being given a list of equations and example problems.

For that matter, this book is a not-so-secret attempt to use the same principle. The hope is that where a simple recitation of the process of science wouldn't be very convincing, a series of illustrative stories will be more memorable and thus more effective at conveying the fundamentally human nature of science and convincing you that you have an inner scientist to discover. And when you become acquainted with that inner scientist, you'll most likely find that he or she is an accomplished teller of stories.

Chapter 12

WHAT'S GOING TO WORK?
TEAMWORK!

SCIENCE AS COLLABORATION

I n October 2012, I had a vicious cold that left me unable to
function without a bunch of over-the-counter medication. I
would make a brief appearance on campus to do paperwork
and teach my class, then immediately retreat home to bed. After three or four days in a pseudoephedrine haze, on the first
day that I managed to get up and about without a handful of
pills, I went to the gym to play basketball for an hour.

While this does rank as one of the dumber things I did in
2012, I had (what seemed like) a good reason: My friend Todd
was in town that day and was stopping by to play. He had been
a regular player in our game for years before leaving to take
another job, and I didn't want to miss a chance to play ball
with him again.

This reason may not make sense to nonathletes, but most
people who have seriously played team sports will recognize
the motivation. In Chapter 10, we talked about the testing aspect of head-to-head competitions, but this is only part of the
attraction of basketball. The other big factor that draws me to
the gym on my lunch hour two or three times a week is the

team aspect of the game. On a good day, with a good team, a basketball game makes you feel part of something larger than yourself. A good basketball play can be extremely complex, but when everything clicks, it's a kick that you can't get from a more individual sport.

Success in team sports relies on more than just individual excellence. I've played with lots of guys who are very good ball-players in terms of their individual skills, but who I don't want on my team. Being a good teammate requires recognizing the strengths and weaknesses of everybody on the team (yourself included) and playing in a way that makes the best use of the skills and that minimizes the weaknesses. If you're the best outside shooter on your team, you try to set yourself up to take those shots, but if you're on a team with a better shooter, you try to set him or her up instead. This approach also holds for match-ups with the opponents—if one of your teammates is being guarded by somebody who can't stop a particular move, you make sure your teammate gets the ball in the right place to make that move, as often as you can manage it.

Successful team play is a combination of communication and experience. Good basketball players are always talking, letting their teammates know where they are and what's going on— where the ball is, when they're about to run into a screen, when they need defensive help, or when they're available to provide help.* And when you know your teammates well, that communication can become almost subliminal—you can anticipate each other's actions, and *know* right where to throw the ball to set up a teammate or where to cut to get the ball for an open shot.

Developing these cooperation and communications skills is as essential to the game as learning to dribble and shoot.

---

* That's even before you include the fine art of trash talking and banter with other players in a friendly game.

And this communication isn't exclusive to basketball—all team sports require cooperation and communication. Basketball is the only game I still play regularly, but I played soccer and rugby in high school and college, and those also demand constant communication. This is one of the primary reasons we teach children to play team sports—not only do they run around and get exercise, but they also learn to work together with others to achieve a particular goal. Sports reinforce elements of the teamwork that is essential for success in just about any career, including science.

The combination of teamwork and science may seem odd, because another of the many unfortunate images popular culture provides about science is the notion of the scientist as a lone genius, working in isolation. This portrayal has more to do with storytelling economy than scientific reality—it's easier and cheaper to tell a story with a single scientist character than a broad array of people. In reality, though, science is an intensely collaborative endeavor.

This aspect of science is reflected in the structure of academic science, with frequent conferences and colloquia in which scientists discuss recent results. It's also a major feature of the stories scientists tell each other. The history of science includes some cherished tales of great discoveries made in isolation, but there are just as many great stories about breakthroughs made through vigorous give-and-take between collaborators or at workshops. The debates about the philosophical foundations of quantum mechanics between Niels Bohr and Albert Einstein at the 1927 and 1930 Solvay conferences are the stuff of physics legend, and the Shelter Island and Pocono Mountain conferences in 1947–1948 played a crucial role in the development of quantum electrodynamics (Chapter 11). Bohr's institute in Copenhagen was central to the development of quantum mechanics in the 1920s, and the often forceful debates between Bohr and his collaborators and guests were a major driver of that progress.*

Even stories that appear to involve a lone genius often turn out, on closer inspection, to be more collaborative than legend would suggest. As mentioned earlier, Einstein was working as a patent clerk when he first developed the theory of relativity in 1905.† But while a patent office may seem like an isolated place for a scientist, he had a close group of friends with whom he regularly discussed science and philosophy (they jokingly called themselves the Olympia Academy). Einstein always credited these discussions, particularly those with Michele Besso, "the best sounding board in Europe," with helping him shape and refine his theories.

The collaborative nature of science is most prominent in modern experimental science, where it is exceedingly rare for new results to come from a single scientist working alone. Even historical results that are usually reported as the work of a single author have generally involved the assistance of technicians and associates who would, by modern standards, be credited as coauthors.

Like good basketball players, then, successful experimental scientists must work well in teams: dividing complex projects to make the work go more smoothly, passing certain tasks off to collaborators with particular skills, and sharing credit for the

---

* On one occasion, Bohr hounded Erwin Schrödinger on some point to such a degree that Schrödinger became ill and took to his bed. Whereupon Bohr pulled up a chair and continued the argument while Mrs. Bohr served Schrödinger soup. (Werner Heisenberg tells this story to Frijtof Capra in Capra's *Uncommon Wisdom: Conversations with Remarkable People* [New York: Simon and Schuster, 1988].)

† Einstein had tried and failed to get a job teaching at a university. This wasn't much easier in 1905 than today, and he had to contend with both a not-terribly-distinguished academic record and an appalling degree of anti-Semitism.

results. While many scientific collaborations, particularly in my field of low-energy atomic physics, involve about as many people as a basketball team, the detector collaborations at the Large Hadron Collider (LHC) involve thousands of members, all of whom play a role in the success of their experiments. In this chapter, we'll look at how experimental physics is done on both small and large scales, and how they share elements with the sort of teamwork familiar to anyone who plays sports.

BIG AND SMALL SCIENCE

My scientific training is in experimental atomic, molecular, and optical (AMO) physics, specifically the study of atoms cooled to within a few millionths of a degree above absolute zero. Some subfields have labs and budgets smaller than those in AMO physics, but AMO physics is definitely on the small end. The lab at the National Institute of Standards and Technology (NIST) where I did my graduate work was a twenty-by-twenty-foot room, and the lab at Yale where I worked as a postdoc was maybe half that size, though both spaces were packed with lasers and optics. In my six years at NIST, only ten people, total, worked on the experiments I was working on, and at Yale, only five.

In contrast, the experimental collaborations at the LHC at CERN work at the opposite extreme in terms of both energy and size.[‡] The LHC itself is housed in an underground tunnel seventeen miles around beneath the border between France and Switzerland, and the main detectors are the size of office buildings. Despite the name, the Compact Muon Solenoid

---

[‡] As I have never worked in experimental particle physics, this discussion relies heavily on interviews with three members of the CMS collaboration. Any major inaccuracies in the description undoubtedly represent original contributions on my part, not anything they said.

(CMS) detector weighs around 12,500 tons, and the other big detector, ATLAS, is comparable.* Building the LHC was a massive undertaking, requiring around ten years and ten billion dollars; as this is an effort not easily duplicated, the LHC is the only accelerator working at the high-energy frontier of particle physics.† As a result, the CMS and ATLAS detectors are run by collaborations with over three thousand members each—collaborations accounting for most of the world's supply of particle physicists.

In spite of the vast difference in scale, the day-to-day experience of AMO and LHC physicists is not all that different. AMO physics experiments tend to involve small teams—generally two or three people working in the lab at any given time, usually in a group supervised by a more senior *principal investigator* (PI), who is rarely involved in lab work. Specific experiments within a research group will be led by a particular member, determined mostly by individual interest, and changing from one experiment to the next.‡ Most experiments in AMO physics involve intricate manipulations of atoms with lasers and magnetic fields, so the bulk of the lab work comes in getting all the necessary pieces to work at the same time. The setup of an experiment involves coordinating both equipment and people, dividing tasks according to individual expertise. Once an experiment is running smoothly, it will often take

---

* ATLAS stands for "A Toroidal LHC ApparatuS," because particle physicists have stopped even trying to justify cute acronyms.

† There are other accelerators operating at lower energies for more specialized purposes, but since the shutdown of the Tevatron at Fermilab, there are no other accelerators with energies comparable to that of the LHC.

‡ One experiment I did as a grad student was started on a whim, because I didn't feel like doing a particular tedious repair, and ended up as a three-month project when the results turned out to be more interesting than expected.

data late into the night, sometimes with lab members working in shifts. For one study at Yale, we literally ran the experiment around the clock until the apparatus caught fire, after which we would clean up, replace the burned parts, and start again.[§]

At the LHC, there is really only one experimental procedure: Accelerate two beams of protons to 99.9999991 percent the speed of light, then slam them together and measure what comes out. In a collision between two fast-moving protons, some of the energy of their motion will be turned into matter (via the world's most famous equation, $E = mc^2$), and the giant detectors are designed to identify the products of these collisions. Most of the exotic particles of interest to high-energy physicists decay into more ordinary matter too quickly to be directly observed, so scientists must infer their creation by careful tracking of the energy and momentum of the decay products, using more sophisticated descendants of the spark chambers discussed in Chapter 2.

The most exotic particles sought by physicists are extremely unlikely to be produced in any given collision, so the LHC experiments need to observe and record vast numbers of collisions. When everything is working, the LHC and its detectors operate 24/7, collecting data continuously for eight to ten hours at a stretch until the proton beam needs to be reloaded.[¶] During normal operation, the detectors are monitored by

---

[§] We were running some electrical switches close to their limits, and after a few days, they would overload. It was faster and cheaper to clean up after the occasional (very small) fire than to reengineer the system to use bigger switches.

[¶] The depletion of the beam is not because of collisions—the beams consist of bunches of around 100 trillion protons, and each time two bunches cross, only about 20 of the 200 trillion protons involved will collide—but because small imperfections in the apparatus allow protons to leak out over time.

teams of five to ten physicists, a group only slightly larger than that for a typical AMO experiment. These physicists work in shifts around the clock to check that the detector is functioning properly and recording useful data.

The real business of experimental particle physics is not just running accelerators, but sifting data. Even though only a tiny fraction of collisions are worth recording—about 350 per second, out of 50 million per second—the amount of data piled up by the LHC is staggering.* The data set from a year of CMS operation runs to 600 petabytes; at typical compression rates, that's over a million years' worth of MP3 music. Unlike AMO physics, where each experiment produces a unique data set specific to that lab, all three thousand members of the CMS collaboration share the same data set. Identifying new particles of interest is a matter of searching through petabytes of data looking for the records of collisions that produced the right sorts of particles to be the thing you're looking for. Once you identify the right collisions, you look for patterns in the energies and distribution of those particles that reveal the exotic particles produced. If a particular type of collision occurs frequently, producing particles with a specific total energy, that suggests you're creating a new particle whose mass corresponds to that amount of energy.

While the data set is shared by the entire collaboration, the specific searches are conducted by research teams not all that much larger than those in AMO physics.† A typical group

---

* The actual data collection is largely automatic, with software triggers scanning the incoming data to identify the particles produced and only record those of greatest potential interest to CMS physicists.

† While any participating group of researchers is free to initiate any search of the data that they like, in practice there is some coordination of topics within the collaboration, to avoid needless duplication of effort.

at a participating institution might include one PI, a postdoc, and a couple of graduate students. Unlike an AMO experiment, though, where all the members of a team work in the same lab, the global nature of the LHC collaborations requires a good deal more communications and coordination. The team working on a particular analysis may be spread halfway around the world, with some members monitoring the detector at CERN while others analyze data in the United States or Asia. Handing the operation on to the next "shift" may mean passing information to a time zone where people are just starting their day. This widespread arrangement requires a good deal of communications infrastructure—collaboration members make frequent use of Skype and other teleconferencing services. And the LHC collaborations use multiple mirror sites and sophisticated network software to efficiently manage searches of the data set. The modern Internet was created by particle physics—the World Wide Web was invented by Tim Berners-Lee at CERN as a way to share information from a previous generation of accelerator experiments.

The biggest difference between large and small science comes not in the day-to-day operation—both AMO and LHC physics involve regular working groups of about the same size—but in the process of releasing a result. Once the data has been collected and analyzed to test a model, the final step of the scientific process is to share the model with the rest of the world via a publication.

In small AMO experiments, this process is relatively straightforward. The person who took the lead on a given experiment will write a draft paper and share it with the rest of the group members who will be coauthors on the paper. The group members review the draft to make sure that all the claims are justified by the data and are stated as clearly as possible. At NIST, we referred to this process as "paper torture," because it involved scrutinizing and arguing about *every single word* and even about things like the positioning and font size of the labels on data graphs. Paper torture sessions could run an hour or more, and tempers would sometimes get a little short by the end, but it was ultimately very effective at producing clear and convincing reports of our experiments.* Once the research group is satisfied, the paper is sent off to a journal for a further review by anonymous referees chosen from other scientists in the field; these people suggest additional changes and recommend whether to publish the article.

By the time a paper is accepted for publication, it's no longer the work of a single person—everybody from the original author's lab mates to the anonymous referees has contributed to the final product. Nevertheless, in AMO and other "small" fields, the first spot on the author list and thus the bulk of the prestige (papers are usually referred to by author names: "Firstauthor et al. found that . . .") goes to the person who took the lead on the experiment and wrote the first draft. The last spot typically goes to the PI; the other coauthors are arranged between them in some mutually agreed-upon order.

While a specific analysis of LHC data may be initiated by a single small team, published results include all three thousand-plus members of the CMS collaboration as authors. It would

---

* This process also helped make me a better writer. As an undergraduate, I could usually coast through classes handing in my first drafts of papers, but paper torture turned me into a dedicated reviser of text. I'll do multiple revisions of a five-sentence email to avoid criticisms from imaginary internal versions of my colleagues from NIST. This doesn't stop me from adding pointless words—I have a terrible habit of putting the word *really* in every other sentence I write—but most of them get removed in the second or third pass and don't end up in the final draft.

obviously be impractical to conduct a single set of paper-torture meetings with all those people, so releasing a result from an LHC collaboration involves a more formal and bureaucratic process.

The specific topics that might be studied by a particular detector are organized into a set of physics groups of a few hundred physicists who share a particular research interest. For CMS, these include searches for the Higgs boson; studies of the top quark; and studies of *exotica*, particles not fitting within the Standard Model. These groups are further divided into subgroups looking at particular ways of identifying the target particles—a subgroup looking at Higgs bosons in conjunction with top quarks, and another looking at Higgs bosons together with muons, and so on. The results of a new analysis are first discussed within a local team (scientists at a particular institution or a small group of collaborators), then presented to a physics subgroup to be checked.

By definition, a subgroup consists of other physicists who have highly relevant expertise, can evaluate the analysis in detail, and can suggest additional tests to apply. If the subgroup approves, the initial authors will write up a description of their analysis and are assigned an analysis review committee of three or four people, including some outside the immediate group. These people are chosen according to their relevant expertise—a team looking for a hypothetical dark matter particle that decays into a top quark and a tau particle would present their analysis to the exotica group and be assigned a reviewer who is an expert in top quarks, another who is an expert on taus, and so on.

The analysis review committee undertakes an exhaustive examination of the results; during this examination, the authors must explain and justify their approach in great detail. They need to convince the reviewers of the validity of the approach in general (why look at this particular collection of collision products?) and of the steps taken to obtain the data (why limit the search to this particular range of energies? why include these collisions, and why exclude those ones?). The analysis review culminates in a presentation to the full physics group, with comments and questions from everyone who attends.

The production of an article for a journal is a similar process, with multiple layers of review, culminating in a collaboration-wide review phase where all 3,400 of the collaboration members are allowed to comment.* Once all the comments and concerns have been addressed—a process requiring multiple iterations and occasionally involving reconciling suggestions that directly contradict each other—the final article is "signed" by the collaboration members who choose to put their name on it, and released as the product of the entire collaboration.

As an outsider to particle physics, I find the idea that credit for the final product is not attached to the specific people who did the analysis the most striking difference between the process I know and the process in the large collaboration. Given the complexity of the experiments, though, this more-inclusive authorship makes sense. Building the detector and starting the experiment is a long and collaborative process, and by the end, as Nick Hadley, the chair of the US CMS collaboration board put it, "everyone knows what measurements are important to do when data arrives, since you have been talking about them for so long." A large fraction of the collaboration members

---

* In principle, at least. In practice, the number of members who actually comment on a paper at some point in the process is more like a few hundred.

could explain in general terms how to approach a particular search, making it difficult to assign greater importance to the initial analysis and writing than, say, the construction and operation of the detector and the computer systems that made that analysis possible.[†] It's easier by far to share credit equally among all signing authors (the lists are generally strictly alphabetical) and avoid what could easily become an endless debate about the precise apportioning of credit.[‡]

The process of releasing a result from a three-thousand-member collaboration may seem protracted to the point of agony. At the level of the individual steps, though, it's surprisingly similar to what we do in AMO physics, only with a thousand times as many authors. And just as the paper-torture process makes the output of a small group better, the more structured review in a large collaboration helps insure high-quality output. All of the CMS physicists I spoke with described it as a very positive process aimed at producing the best physics possible.

At both the largest and the smallest scales, teamwork is essential to physics. Whether the research team fits in a single office or is spread halfway around the world, setting up experiments and collecting data requires coordinating and sharing tasks with colleagues. Whether the author list is the size of a basketball team or a small town, producing a published paper requires the input of many others. The essential elements of

---

[†] Obviously, everyone couldn't predict the result or complications along the way, but most people in the collaboration could outline the basic process.

[‡] The members of the collaboration know who initiated a particular search and can share this information in job applications and letters of reference, but for general public presentations, all final results are the work of the entire collaboration.

communication and cooperation should be familiar to anyone who has played a team sport; whether on the field or in the lab, success demands that you play well with others.

OTHER TYPES OF COLLABORATION

While the popular image of scientists as eccentric loners is greatly exaggerated, scientific collaborations often need to accommodate some unusual working habits. Julian Schwinger, one of the aforementioned coinventors of QED, was the opposite of a morning person. When teaching at Purdue in the early 1940s, he only reluctantly scheduled undergraduate courses during daylight hours, and the department secretary had standing orders to call him at noon on days when he had to teach, to make sure he was awake.* His preferred routine was to sleep until dinnertime; have a breakfast of steak, French fries, and chocolate ice cream; and then work through the night.

His nocturnal habits presented something of a problem when Schwinger was recruited for the large-scale physics projects of World War II, but he was brilliant enough that people found ways to work around his eccentricities. During the couple of months Schwinger spent working on reactor physics for the Manhattan Project, Bernard Feld worked a sort of swing shift so he could communicate the problems of the day to Schwinger when the night owl arrived. When Schwinger shifted to working on radar technology for the Radiation Laboratory at MIT, he became a one-man night shift in the Theory

---

* Night owls are common in physics. When the great Scottish physicist James Clerk Maxwell arrived at Cambridge he was told there would be a mandatory 6 a.m. church service, and he supposedly replied, "Aye, I suppose I could stay up that late."

Division—colleagues would leave notes on his desk describing problems they were having when they left for the evening and would return in the morning to find neatly worked-out solutions left by Schwinger.[†]

Some more conventionally scheduled scientific collaborations also develop into deeper partnerships. In 1925, Frédéric Joliot started as a student at the Radium Institute run by Marie Curie, where he worked with Curie's daughter Irène, who had just completed her Ph.D. The two hit it off, and they were married in 1926; though Marie Curie was initially skeptical, the marriage turned out to be a great success both personally and scientifically.

Prior to their marriage, both Irène and Frédéric had solid but unspectacular scientific credentials, but when they combined their efforts, the Joliot-Curies produced an impressive string of discoveries and near misses in the early 1930s. In early 1932, they published some intriguing results (following on work by Walther Bothe) involving radiation produced when alpha particles struck lighter elements. The Joliot-Curies didn't appreciate the full significance of these results, but James Chadwick at the Cavendish Laboratory in Cambridge did. He quickly performed a classic series of experiments demonstrating that the new radiation was a neutral particle with the mass of a proton, securing himself the 1935 Nobel Prize for the discovery of the neutron.

In April of that same year, the Joliot-Curies also produced an exceptionally clear set of photographs showing a small

---

[†] A twenty-first-century version of Schwinger would presumably just telecommute. These stories are from Jagdish Mehra and Kimball A. Milton, *Climbing the Mountain: The Scientific Biography of Julian Schwinger* (New York: Oxford University Press, 2000).

number of particles with the mass of electrons whose tracks through the magnetic field in their detector curved in the opposite direction from the other electrons produced. They interpreted these as electrons moving in the opposite direction, apparently created by particles that passed all the way through their detector chamber and struck the wall. Four months later, Carl Anderson at Caltech produced a similar set of photographs with the correct interpretation: These particles were positrons, the antimatter equivalent of electrons. Anderson won the 1936 Nobel Prize for this discovery.*

The third time was the charm for Frédéric and Irène, though. In 1934, they were the first to demonstrate the artificial production of radioactive elements. When they bombarded relatively light atoms with alpha particles, a tiny fraction of the alphas would be absorbed by nuclei in the target, becoming unstable isotopes of different elements. The Joliot-Curies showed that they could make radioactive nitrogen from boron, radioactive phosphorus from aluminum, and radioactive silicon from magnesium. This was a discovery of revolutionary importance—previously, scientists wanting to study radioactive substances had needed to distill them in tiny quantities from naturally occurring samples, as Marie Curie did (Chapter 3). Using the Joliot-Curie discovery, scientists could generate artificial isotopes more efficiently from stable samples. This capability opened up a wider range of substances to detailed study and allowed the production of medically useful isotopes

---

* The positron had previously been predicted by Paul Dirac, so Anderson's discovery also helped secure his 1933 Nobel. Also in 1936, Anderson discovered the muon, a wholly unexpected particle similar to the electron but much heavier, the first of an explosion of new particle discoveries that revolutionized subatomic physics.

on an industrial scale.[†] Their research won Irène and Frédéric a well-deserved Nobel of their own, the 1935 Prize in Chemistry.[‡]

The vast majority of scientific collaborations, of course, operate somewhere between the extremes represented by Julian Schwinger and the Joliot-Curies. Very few scientists marry their coworkers, and fewer still communicate with their colleagues exclusively through notes left on desks. Whatever form it takes, though, teamwork is as essential to scientific progress as it is to basketball.

MORE SPORTS

My favorite sport to watch as a fan is NCAA (National Collegiate Athletic Association) basketball, because the game as played at the college level is closer to what I'm familiar with as

---

[†] This is also the process Luis Alvarez later used in discovering the asteroid impact that killed the dinosaurs (Chapter 2).

[‡] They went on to lead fascinating lives after these discoveries, as well. Their studies of artificial radioactivity provided information and inspiration for studies of uranium fission that led to the wartime atomic bomb projects in the United States and Germany. When the Nazis came to power and war appeared imminent, the Joliot-Curies sealed all their results on artificial radioactivity in the vault of the Académie des sciences, where the documents remained until 1949. When France fell in 1940, Irène was in Switzerland convalescing from tuberculosis; during the war years, she made several trips into occupied France to visit her family, enduring German detention at the border. Frédéric remained in Paris, ostensibly working on atomic physics, but used his research as cover for manufacturing radios and bomb-making materials for the French Resistance; for this, he was named a commander of the Legion of Honor with a military title and awarded the Croix de Guerre. After the war, they both played key administrative roles in the French scientific establishment, before losing their positions at the height of the cold war because of their ardent socialist views.

a player; a number of small rule changes and cultural norms around NBA (National Basketball Association) basketball make it less appealing to me. Of course, for many players, college basketball is merely a brief stop on the way to an NBA career; NBA rules require a player to either be nineteen years old or have spent one year in college before he enters the NBA draft. Many of the very best players now go to college for only a single year before moving on to the NBA.

When this rule was put in place in 2005, there was a lot of hand-wringing from college basketball fans about how "one-and-done" players were going to ruin the game. Unscrupulous coaches, they felt, would use underhanded means to attract several of these players and put together an unbeatable team of mercenaries, tainting the whole idea of college athletics. The 2012 NCAA basketball champions, the Kentucky Wildcats, are an example of such a team: Their coach, John Calipari, had left his two previous jobs just before major NCAA rules violations were uncovered, and after the season, five of the school's top players, including three freshmen, left school for the NBA draft.

Other than that one Kentucky team, however, one-and-done players have not had the devastating impact they were supposed to. Even the 2012–2013 Kentucky team, which replaced its five departing stars with four new star recruits (two of whom would leave after only one year), couldn't repeat the success: Although the team started the season ranked number three in the country, it soon dropped out of the rankings altogether and lost in the first round of both its conference tournament and the NCAA tournament. In fact, only one other NCAA champion since 2000 has featured a one-and-done player, the 2003 Syracuse Orange, whose star forward was Carmelo Anthony.

The dire predictions about one-and-done players haven't come true because while basketball is the easiest sport in which a single great player can have significant effect, it's still a team game, and success requires all five players on the court to work together as a team. What's more, effective teamwork requires the right mix of players and personalities and often takes time to develop. In the same period, meanwhile, the NCAA tournament has seen improbable runs by teams with much less talent (on paper) but with great team chemistry. Consider, among others, the Final Four appearances by George Mason in 2006 and Virginia Commonwealth in 2011 and consecutive championship game appearances by Butler in 2010 and 2011.

While the vast majority of basketball players will never experience those heights, these "Cinderella" teams are an extreme example of the team experience that draws many people to sports. With the right mix of people, you can accomplish far more than would be expected considering the individual abilities of the team.* Even at the recreational level, experience can top athleticism—at our lunchtime pickup games, we'll occasionally play "old guys versus young guys," pitting the faculty against students who are twenty or more years younger. We old guys win a lot of these games, in large part because we've been playing together for years and know how to work as a team.

The same thing is true of any activity involving more than one person: Success almost always requires working together

---

* And of course, the wrong mix of people can be a disaster even if their individual abilities seem to predict greatness. Sports fans can point to any number of cases where adding a great individual player to a good team made everything worse rather than better; a good example might be the 2012–2013 Los Angeles Lakers, who added All-Star players Steve Nash and Dwight Howard to an already loaded roster, but struggled with team chemistry all season, barely reaching the playoffs, where they were swept in the first round. Scientific collaborations can also fail, sometimes in spectacular ways that result in endless recriminations and even lawsuits.

with others, which means finding ways to communicate and coordinate activities among a group. This means both finding the right people to work with and setting up structures to allow them to succeed, whether it's a formal structure like that used for the CMS collaboration or an informal arrangement like the note-passing used to accommodate Schwinger's night-owl tendencies. Combining the efforts of your inner scientist with those of others may require some ingenuity, but the results are well worth the extra effort.

As for my ill-advised return to the court, mentioned at the start of this chapter, after dragging myself to the gym for the chance to play with my friend Todd, we spent most of the day in different games. I only got to play one game on Todd's team, but as I had hoped when I packed to go to the gym that morning, it finally paid off. On one fast break, another teammate put up a shot as I was running down the foul lane, with Todd not far behind. There was a defender in position to fight me for the rebound, but rather than trying to catch the ball, I just batted it out toward the corner. Todd was there to catch it, as I had known he would be. As the defender ran out to pick him up, I backpedaled furiously under the rim, where I caught the pass I knew would be coming and put in an easy layup. That one brief sequence by itself was worth dragging myself to the gym without medication. That kind of teamwork and communication is a hard-to-beat experience, whether on the court or in the lab.

Chapter 13

TALKING SPORTS

When I was growing up, my father would bring home a daily paper and consume (often with great gusto) the baseball box scores. There they were, to me as dry as dust, with obscure abbreviations (W, SS, K, W-L, AB, RBI), but they spoke to him. Newspapers everywhere printed them. I figured maybe they weren't too hard for me. Eventually I got caught up in the world of baseball statistics. (I know it helped me in learning decimals. . . .)

Or take a look at the financial pages. Any introductory material? Explanatory footnotes? Definitions of abbreviations? Almost none. It's sink or swim. Look at those acres of statistics! Yet people voluntarily read the stuff. It's not beyond their ability. It's only a matter of motivation. Why can't we do the same with math, science, and technology?

—Carl Sagan, *The Demon-Haunted World*

Sharing results with other scientists is absolutely critical to scientific progress and demands that information be conveyed completely and accurately. As with all human communications, though, there is great potential for errors and misunderstandings when people disseminate scientific results. Because of the need to exchange complete descriptions efficiently, most sciences have developed an extensive jargon with very precise, technical meanings known to experts in the field.* More than that, though, science relies heavily on mathematics, as mathematical formulas can be perfectly unambiguous in a way that words can never match.

The combination of technical jargon and mathematical notation can make science very off-putting. Many nonscientists will throw up their hands when confronted with jargon terms, and guides to science writing are full of dire warnings to avoid jargon. The reaction to math is even more extreme— one nonscientist who read a draft of my first book reported becoming physically angry at running across an equation in one chapter.[†] An old joke in popular physics writing claims that each equation appearing in a book cuts its sales in half.[‡]

Both of these criticisms are somewhat unfair. Scientists are hardly the only professionals to use specialized jargon— any group of people doing any job more complicated than

---

[*] These terms sometimes overlap with everyday words in a way that is confusing to people outside the field. To use a simple example from physics, in normal conversation, there's little distinction between the terms *speed* and *velocity*, both of which refer to how fast something is going. For physicists, though, there is a critical distinction between the two: Speed is a measure only of how fast an object is moving, whereas velocity is a measure of how fast an object is moving *and in what direction it's moving*. A car moving at 60 mph northward has the same speed as one moving at 60 mph southward, but the two have very different velocities.

[†] The book was *How to Teach Physics to Your Dog* (New York: Scribner, 2009), which explains quantum physics for nonscientists through imaginary conversations with our German shepherd mix, Emmy.

[‡] This quip is often attributed to Peter Guzzardi, the editor of Stephen Hawking's *A Brief History of Time: From the Big Bang to Black Holes* (New York: Bantam Books, 1988). That book contains one equation and has sold more than ten million copies worldwide.

digging holes and filling them back in will eventually develop shorthand terms to communicate more effectively. Even journalists who issue warnings about jargon in science stories use technical terms when discussing their own craft—*lede*, *graf*, *hed*, and other words that look like misspellings to outsiders.

The incomprehensibility of math is also overstated. Not only have books making heavy use of math sold well—*The Theoretical Minimum* by Leonard Susskind and George Hrabovsky is chock full of equations and reached number thirteen on the *New York Times* best-seller list—but nonscientists use math to a greater degree than conventional wisdom would have us believe. In fact, you can find surprisingly sophisticated mathematical techniques being employed by the very people pop culture would have us believe are the least able to handle them: athletes and sports fans. You're more likely to hear quantitative data invoked to settle arguments on ESPN than news networks like CNN.

In fact, while the core of sports fandom remains watching the actual contests, in the last several years, math has become more central to the sports experience. Fantasy sports have become a huge business, with some thirty-five million people participating in some sort of organized fantasy league in 2012, according to the Fantasy Sports Trade Association.§ These leagues are just numbers on paper—fantasy sports "owners" select a list of players in the appropriate sport to be on their imaginary team, and the owners accrue points according to the statistics accumulated by those players in the actual games. In fantasy football, for example, a team might get one point for every ten passing yards by the quarterback. In other systems,

---

§ The existence of the trade association itself is also a great demonstration of the popularity of fantasy sports.

such as Rotisserie League Baseball, award points are based
on rank within the league, so in a ten-team league, the owner
whose team accumulates the highest total number of runs over
the season gets ten points for that category, the second-highest
nine, and so on.

As a scientist interested in the public understanding of science, I find fantasy sports fascinating because they're entirely
based on math. For as much as we hear that ordinary people
don't like or don't understand math, tens of millions of people
devote large amounts of time and energy to games that are
entirely based on statistics. Fantasy wins and losses are accumulated through the manipulation of numbers, and many
decisions about which players to start and which to trade are
based on statistical estimates and projections.

Paralleling the rise in popularity of fantasy sports, there
has also been an increase in the sophistication of the methods
used by teams, fans, and sports pundits to track and evaluate
player performance.* Thirty years ago, sports fans were primarily interested in simple cumulative stats—points scored,
runs batted in, total rushing yards. While those are still tracked
and cited, new and more complicated measures have become
popular and are frequently invoked in sports arguments: slugging percentage, offensive efficiency, value over replacement
player, and so on. Many of these factors are quite complicated
to calculate, but they're used readily by people who, stereotypes say, shouldn't understand math. But as Carl Sagan notes
at the beginning of this chapter, nonscientists are perfectly capable of understanding this stuff when they care to.

As with science, the use of statistics in sports provides
a common language enabling fans to share complete and

---

* Both of these are driven by a common factor, namely, the continued increase of computing power.

accurate information. Many of the techniques used by sports fans are closely related to mathematical tools that scientists use to share quantitative results. Anyone who can manage a fantasy sports team or follow an argument on ESPN can use the same tricks and techniques to understand a wide range of scientific results. In this chapter, we'll look at some areas where sports and science share statistical techniques, and how fantasy sports owners are already using their own inner scientists.

NORMALIZING THE COURT: ADVANCED BASKETBALL STATS

One of the fundamental issues facing experimental scientists trying to share results is that every lab is a little bit different. No two particle detectors in physics are exactly the same, no two chemical samples are prepared by perfectly identical processes, and no two organisms have the same life histories. Even when the same apparatus is used to repeat a measurement, there can be changes over time, sometimes for the worse (parts begin to wear out and break down) or sometimes for the better (scientists and technicians are forever making small tweaks and upgrades to their apparatus). Whether for the better or for the worse, the changes affect the exact conditions of a particular measurement.[†]

For this reason, scientists go to great lengths to measure and compare universal quantities. Rather than talking about the absolute number of particles detected at an accelerator, physicists talk about the *rate* of production, which is the number of particles detected divided by the number of collisions

---

[†] Even theoretical calculations can be affected by some random errors and uncertainties, as a result of small differences in computer hardware and software used to simulate complex systems.

needed to produce them. This measurement, usually expressed as a probability of producing a particle in a given collision, removes details that can easily vary from one lab to the next and allows more general comparisons. A physicist at a different accelerator can divide the number of particles detected at the second accelerator by the number of collisions that they measured and get a rate that ought to be the same, even if the total numbers of particles and collisions are very different.

Similar issues come up in sports. A fundamental problem for people running teams—real coaches and managers or fantasy owners—is to compare two players and decide who they want on their team. As with scientific measurements, though, this comparison is complicated by the fact that no two sporting events are perfectly identical. A whole host of factors changes from one game to the next—injuries, weather, opponents, and so forth. The job of a statistician trying to evaluate athletes is to find universal quantities that can be fairly compared across different conditions, quantities that reflect the underlying skill of the player, not mere luck.

In the case of professional baseball, where each team plays 162 games and faces every other team in its league several times in a season, simple averaging can suffice.* Given enough games against the same opponents, the factors that vary from one game to the next mostly cancel each other out. Better-than-average performance in one game is balanced by worse-than-average performance in another. Looking at how two players compare in their performance against a common opponent over multiple games gives a reasonable basis for comparison.

---

* Each team plays nineteen games against each of the other four teams in its own division, and six or seven games against the other teams in the same league, plus a handful of interleague games.

If you try to compare players across leagues, however, the problem becomes much more difficult. When the players you would like to compare don't face any common opponents or only play them a few times, valid comparisons are much harder to make. This problem is particularly acute when you are looking at my favorite sport as a fan, college basketball. Almost 350 colleges and universities play basketball at the Division I level in the NCAA, and by rule, teams are limited to around thirty regular-season games. Most teams play conference opponents twice, so any given team plays maybe twenty different teams during the season, about 6 percent of the total number in the NCAA. For many pairs of teams, it will be impossible to find *any* common opponents that could serve as a reference for comparisons between players.

Even beyond the lack of common opponents, college basketball features a wide array of styles of offense and defense, which fundamentally change the way their games play out. Some teams play an "up-tempo" game, playing aggressive defense to force turnovers or quick shots and trying to score quickly on offense—Rick Pitino's teams at Kentucky and Louisville are good examples. Other teams prefer to play at a more deliberate pace, waiting as long as it takes to get the shot that they want—Princeton and Wisconsin are famous as slow-playing teams. While these differences in style are part of the attraction of college basketball, they make life very difficult for anyone trying to compare players.[†] According to statistics from college basketball analyst Ken Pomeroy, the fastest-playing teams in college basketball average about seventy possessions per game, while the slowest average about

---

[†] As opposed to the NBA, which tends to be a bit more uniform, for a variety of reasons.

sixty, and the game-to-game variation can be much greater. A player on a high-tempo team thus has up to ten more chances to score per game than a player on a low-tempo team.* For a good offensive player, that can mean an extra two points per game, or about sixty total points over the full season. Likewise, players who suffer an injury or who spend time on the bench for whatever reason will have fewer opportunities to accumulate points and rebounds.

Given this variability, it doesn't make sense to pay too much attention to absolute numbers—the total number of points, rebounds, blocked shots, and so forth accumulated over a season.† These numbers are easily affected by the pace of play and other factors outside the control of an individual player. A better measurement of individual contributions comes from *tempo-free stats*, which generally take the form of percentages and probabilities. Serious sports statisticians rely on measures like points per possession (fairly self-explanatory), shooting percentages (which measure the probability that a player will score when attempting a shot), rebounding percentages (the number of rebounds a player grabbed divided by the number of missed shots while he was in the game), and blocked-shot percentages (the number of shots a player blocked divided by the number of shots the opponent attempted). Like the numbers reported by most scientific studies, these should remove factors that are beyond the control of the individual player and provide a more consistent measure of individual ability.

---

* The difference is somewhat smaller for the NBA, with the fastest teams in the 2012–2013 season averaging ninety-eight possessions a game, and the slowest about ninety-one.

† These totals are still used for some annual awards, but people evaluating players don't pay them much attention.

While the problem of game-to-game variability is most acute in college basketball, all of the major (American) sports make use of these kinds of statistics. Baseball fans track batting averages (more about these shortly), which express the probability of a player getting a hit when he steps up to the plate. Football fans track statistics like completion percentages for quarterbacks and yards per carry for running backs. Just like scientists comparing results between labs, sports fans and fantasy owners deal primarily in universal quantities that can be fairly compared across the entire league.

BASEBALL: AN ARRAY OF AVERAGES

Although universal statistics provide a common language for sharing information about the performance of players and teams, the use of statistics does not ensure complete agreement about which numbers to use. Just as two English speakers can disagree about the proper word choice for describing the same objects, both scientists and sports fans have vigorous arguments about which universal quantities are the most useful for describing a given situation. These differences are nicely illustrated by the array of statistics used to measure batting performance in baseball.

By rule, a complete baseball game will involve at least twenty-four times that a player comes to bat for each team.[‡] Each of the nine players on the team can thus expect to come up to bat around three times per game, and there are 162 games in the season, so a player who's with a given team for

---

[‡] Nine innings with three outs each is twenty-seven, but if the home team is ahead at the middle of the ninth inning, it doesn't bat, so the minimum number of batters is twenty-four.

the full season can expect to bat several hundred times. Each of those trips up to bat can be thought of as an experiment to measure that player's hitting ability and offensive value to his team. Baseball thus provides a very rich source of data, going back over a hundred years, that sports statisticians can investigate for the best ways to measure individual performance. But there are many ways to combine these measurements beyond the simplest "add all the values and divide by the total number of measurements" form of averaging.

The most straightforward question to ask is this: What is the chance that when a given player comes up to bat, he *doesn't* make an out? The resulting statistic is the *on-base percentage.** This number consists of the sum of all the ways a batter can reach base without making an out (getting a hit, being walked, or being hit by a pitch) divided by the number of plate appearances for that batter. This is the most basic measure of offensive competence and the analogue of the simplest add-and-divide average. Let's look at two examples.† Over his thirteen-year career with the Mariners and Yankees, Ichiro Suzuki stepped up to the plate 9,278 times and reached base 3,350 times for a career on-base percentage of .361. Another great player with a very different style, Barry Bonds, stepped up to the plate 12,606 times over twenty-two years, reaching base 5,597 times for a career on-base percentage of .444. By this measure, then, Bonds is the better offensive player.

The simple add-and-divide method isn't always appropriate, though. For some individual scientific measurements,

---

* Baseball's flair for naming things is comparable to that of most scientists.
† These two were chosen as players famous enough to be familiar to those (like me) who only follow baseball casually and as players who are successful through very different approaches.

there is often a good reason to suspect that something may have thrown the result off. In that case, it is appropriate to exclude those measurements from the average, as they don't accurately reflect the underlying value of the property being measured. Trimming out the suspect data can better reveal the underlying phenomenon of interest.

The analogous baseball statistic is the batting average. While there are three ways for a batter to reach base without making an out, two of those—being walked and being hit by a pitch—arguably say more about the pitcher (in a bad way) than about the ability of the batter. The batting average divides the total number of hits recorded by a particular batter by the number of *at bats*, the number of plate appearances minus those where the batter was walked, hit by a pitch, or credited with a sacrifice (an out that advanced a runner already on base). The career batting averages of .310 and .298, respectively, for Suzuki and Bonds reflect that Bonds was walked many more times—about 140 times per season, compared with 40 for Suzuki. If you're only interested in the ability to hit the ball, this would suggest Suzuki is the better of the two.

Another issue in baseball is that not all hits are the same. A small but particularly speedy runner, like Suzuki, may rack up a lot of hits by narrowly outrunning throws to first base, while a larger, slower player like Bonds (the iconic example of a power hitter from the "steroid era" of baseball) might hit a lot of home runs, but get very few singles. Those two types of players might end up with similar batting averages, but have very different impacts on their respective teams.

The statistic that attempts to account for the impact of different types of hits is the *slugging percentage*, an example of what's known in science as a *weighted average*. To calculate the slugging percentage, you add up the number of singles, two times the number of doubles, three times the number of triples, and four times the number of home runs and divide that by the number of at bats. This gives a number that is generally higher than either batting average or on-base percentage and reflects the greater impact of extra-base hits. This statistic most clearly shows the contrast in style between Bonds and Suzuki—Suzuki's career slugging percentage is .414, while Bonds's is .607. When Bonds hit the ball, it tended to go a long way, whereas Suzuki made it to first ahead of the ball on a lot of short hits that would have been outs for slower players.

Weighted averages of this type generally occur in science when combining results from measurements made by different labs or using different techniques. Different measurements will have different intrinsic uncertainties (more about this in the next chapter), and when putting the results together, scientists will assign a greater weight to measurements whose intrinsic uncertainty is smaller. Like home runs in baseball, high-precision measurements are more valuable than low-precision measurements and are thus given a greater emphasis in single overall values that combine the results of multiple techniques. In physics, for example, the official values of the fundamental physical constants (including both particle properties like the mass of an electron and universal constants like Newton's gravitational constant) are published periodically by the international Committee on Data for Science and Technology (CODATA). The CODATA values are a weighted average of the published experimental values; the actual process of weighting and averaging is much more complicated, but the concept is the same as the slugging percentage in baseball.

Trimming and weighting data sets are common practices in science, but slightly risky. For scientists who know in advance the answer they expect (or hope) to get, the temptation to simply exclude a few points here and maybe tweak the weighting there to match the desired result can be too much to resist. Many scientific controversies turn on which data points were excluded and how that affects the data. Even some results that have held up—in physics, Robert Millikan's measurement of the electron charge, and in genetics, Gregor Mendel's experiments with breeding pea plants—involve data selection that some claim goes beyond modern standards. To avoid biasing their results, many high-profile experiments in physics now use *blinded* data analysis, where the true value is hidden by, for example, adding a large offset known only to the computer that shifts the average value away from any expected value the researchers may have in mind as a target. The human scientists in charge of the analysis make all their decisions on what to include and exclude while looking at the data including the offset and only reveal the true result after those decisions are made.

Issues of data selection and weighting are, of course, the reason for the existence of three different batting statistics in baseball. Batting average has the longest history, officially tracked since the late 1800s, but there have been arguments about whether other measures might be better since at least the 1950s. Official tracking of on-base percentage began in 1984, and most statistical services now report all three statistics, letting fans and fantasy baseball owners choose whatever they prefer.* Both Bonds and Suzuki are great players—both would be headed for the Hall of Fame were it not for Bonds's drug issues—but if you were putting together a fantasy baseball team and your scoring system gave more weight to batting average than power hitting, you might reasonably prefer Suzuki.

---

* Or a combination—lots of fans and teams use the sum of on-base and slugging percentages as a single number to measure batting.

PROJECTION METHODS: PAST PERFORMANCE
DOES NOT GUARANTEE FUTURE RESULTS

The ultimate goal of communication for both scientists and sports fans is to inform and coordinate action of some kind. Scientists share models to enable independent replication, and so new discoveries can be used to plan future measurements. Fantasy sports fans share player statistics to allow better decisions about which players to play or to acquire through drafts or trades.

Broadly speaking, both groups share a common goal, namely, to predict the future. Given a bunch of data about the past—whether scientific measurements or sports box scores—we would like to construct a model that allows us to predict the outcome of future events, given some basic information. In either profession, the rewards for such a model are great: A model that accurately predicts the outcomes of sporting events could make its creator a fortune placing bets in Las Vegas, while a scientific model that can accurately predict future measurements could earn its creator major research grants and tenure.

Prediction models come in all sorts of forms, some of them dizzyingly complex. The basic process used to generate and test such a model, however, is the same for all of them and can be illustrated using one of the simplest such tools around: the *Pythagorean winning percentage*. Introduced by legendary baseball statistician Bill James, this formula purports to predict the winning percentage of a baseball team over the course of a season using only two factors: the number of runs a team scores and the number of runs scored by their opponents. Given the scoring for the first half of the season, for example, you can use this percentage to predict the expected number of wins in the second half, with more success than simply extrapolating from the midseason win-loss totals.

There's an obvious appeal to the idea, as runs scored and runs against are, ultimately, the two factors that matter: By definition, the team that wins a game has scored more runs than its opponent. The name Pythagorean comes from the form of the original formula:

$$Win\% = \frac{(Runs\ For)^2}{(Runs\ For)^2 + (Runs\ Against)^2}$$

Squaring the run totals is reminiscent of the famous Pythagorean theorem in geometry, relating the lengths of the sides of a right triangle by the formula $a^2 + b^2 = c^2$.

James published this in the late 1970s as part of his series of groundbreaking books about baseball statistics. The Pythagorean winning percentage began as an empirical relationship—that is, in playing around with scoring statistics for baseball, he found that this formula did a pretty good job of matching the actual winning percentage of most teams in baseball history. Numerous articles have since been written showing how this formula can be derived mathematically starting with some simple assumptions.

As computing technology has improved, however, checking and testing these relationships has become easier, allowing refinements of the basic formula. These days, conventional wisdom in baseball is that a slight modification of the formula does a better job predicting reality:

$$Win\% = \frac{(Runs\ For)^{1.82}}{(Runs\ For)^{1.82} + (Runs\ Against)^{1.82}}$$

The modified version reduces the exponent slightly, from 2 to 1.82.*

Given the demonstrated success of James's formula for baseball, statisticians for numerous other sports have attempted to define their own "Pythagorean theorems," but this is where things depart significantly from the Pythagorean analogy. For professional football, the commonly used formula is

$$Win\% = \frac{(Points\ For)^{2.37}}{(Points\ For)^{2.37} + (Points\ Against)^{2.37}}$$

For college basketball, Ken Pomeroy uses

$$Win\% = \frac{(Points\ For)^{10.25}}{(Points\ For)^{10.25} + (Points\ Against)^{10.25}}$$

And for NBA basketball, both

$$Win\% = \frac{(Points\ For)^{13.91}}{(Points\ For)^{13.91} + (Points\ Against)^{13.91}}$$

and

$$Win\% = \frac{(Points\ For)^{16.5}}{(Points\ For)^{16.5} + (Points\ Against)^{16.5}}$$

---

* Unlike squaring a number, where you simply multiply it by itself, a fractional exponent isn't something you can easily do with pencil and paper. It's a simple operation for a calculator or a spreadsheet, though.

are used by different statisticians (with lots of arguments about which is better, turning on issues of data selection and weighting, as discussed in the previous section).

Where do all these strange numbers come from? They're generated by a trial-and-error process that is virtually identical to the fitting process used to test scientific models. It's something you can easily reproduce in Excel if you'd like to validate the Pythagorean formulas above. Win-loss records and scoring statistics are readily available from numerous sites on the Internet, and the Pythagorean formulas are easy to enter into a spreadsheet. The formulas give a "predicted" number of wins that can be compared with the actual team records. Then it's just a matter of trying different exponents to find which one best matches the actual records. Figure 13.1, for example, shows the performance of the Pythagorean model with different exponents between 10 and 20 for the 2012–2013 NBA season.[†] On the graph, "win difference" means the difference between wins predicted by the Pythagorean formula and the actual wins.

You can see right away that there's a "best" value for the exponent, where the difference between the predicted and actual wins is a minimum, at around 13.75. Sports statisticians will test their models over many seasons' worth of data, and the numbers in the preceding formulas represent the results of such tests on historical data.

The predicted records from these methods give a baseline measure of what we "ought" to expect a team's performance to be. These provide a basis for assessing the role of luck in a team's performance—a team that won significantly

---

[†] Data from ESPN, "Win Percentage" Web page, http://espn.go.com/nba/stats/rpi/_/sort/PCT.

Figure 13.1. Difference in total wins between actual 2012–2013 NBA results and the Pythagorean prediction thereof, for various values of the exponent in the Pythagorean formula.

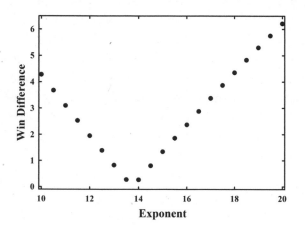

more games than the Pythagorean prediction got "lucky," whereas one with more losses than expected was unlucky. These figures also provide a simple way of predicting the future—a team well above or well below the expected winning percentage at midseason can be expected to move back toward the prediction, and a team that significantly underperformed its expectation in one season will most likely improve the next season, even without drastic changes in players or coaches.

These predicted records can also be used as a crude predictor of individual games. If we regard the projected winning percentage as a measure of a team's quality, another Bill James invention, the *log5 probability*, gives the probability of a team's defeating a particular opponent in terms of the winning percentages of the two teams.* Using the NBA data above, for example, log5 predicts that the Miami Heat (with a projected winning percentage of 75.3) ought to beat the San Antonio Spurs (71.1 percent) about 55.4 percent of the time, or 3.9 games out of 7. The Heat in fact defeated the Spurs in the 2013 NBA Finals 4 games to 3, consistent with the model's predictions.

Similar techniques abound in science. Given a large set of data from past experiments, scientists will come up with a mathematically plausible model to describe it and then use this sort of fitting process to find the best values for parameters in the models.† The scientists report these fit parameters to other scientists, who can check the parameters against their own data and use them to predict the outcomes of future measurements.

This evaluation process is also iterative—any successful model will be tweaked and refined, becoming more complex but hopefully also more accurate. The Pythagorean and log5 methods are among the simplest prediction systems around, relying solely on scoring; hard-core modelers will use dozens of different parameters, attempting to capture the effects of location, injury, and all the other little things that can tip a game one way or another. Similarly, many scientific studies (particularly in social and behavioral science) will use *multivariate*

---

* The log5 formula predicting the winning percentage of team A over team B is $Win\% = \frac{A\% - (A\%)(B\%)}{A\% + B\% - 2(A\%)(B\%)}$, where $A\%$ and $B\%$ are the Pythagorean win percentages for team A and team B.

† The fitting process is generally automated by a computer. If you've ever had Excel add a "trendline" to a chart, this is what it's doing: trying different values for the parameters describing the line and finding the ones that work best. Professional scientific analysis software is much more sophisticated, but essentially doing the same thing.

regressions looking at the effects of dozens of experimental and environmental parameters. Conceptually, though, the process is the same as for the Pythagorean formulas above and ought to be comprehensible to any sports fan.

Statistical methods for comparing players and communicating those evaluations to others have transformed sports fandom in the last couple of decades. The mathematical sophistication of sports media coverage has increased dramatically, and the rising popularity of fantasy sports has spread those methods well beyond team offices and TV studios. The statistics-driven transformation also extends to the teams themselves, with most franchises employing statisticians to invent new methods for measuring and predicting performance in hopes of gaining an edge.

If anything, this transformation seems likely to accelerate in the next few years. The NBA is investing millions of dollars outfitting its arenas with elaborate multicamera systems to allow the tracking of each player's actions at an unprecedented level of detail. This trove of information will undoubtedly lead to new ways of analyzing the flow of a basketball game and the performance of individual players.

The increasing reliance on number-crunching management methods bothers some traditional fans, but the methods have undeniably proven effective. Objective mathematical techniques have allowed canny management to exceed expectations, as dramatized (and somewhat mythologized) in the book and film *Moneyball*. And the great power and reliability of statistical methods are not limited to sports. Statistical projections are behind all sorts of everyday practices, from the purchase-tracking routines that credit card companies use to detect possible fraud to the recommendation engines used by Internet companies like Facebook and Twitter. For example, if you normally use your credit card only to buy gas and groceries at local stores, and then it's suddenly used to purchase a canoe in Buenos Aires, you can expect a phone call from your bank. Or if somebody whose posts you read and like tends to read and like posts from a third party, chances are good that you will enjoy these posts as well, and Facebook or Twitter will suggest this new contact. Statistics also offer a popular and reliable tool for understanding politics, as we'll see in the next chapter.

FANTASY REDUX

As a passionate fan of specific teams—on several occasions in grad school, I woke my housemates up by yelling at games on TV—I've always found fantasy sports a little odd.* My primary interest as a fan is in seeing the teams I root for win, and I dislike being in a position where I "have" to root for a player on a rival team to do well. When I enter office pools for the NCAA basketball tournament in March, I always pick my favorite teams to go to the Final Four and predict early losses for the teams I dislike, so as to keep my rooting interests and my betting interests aligned as much as possible.

When I started working on this book, though, I knew I needed to say something about sports statistics, so I joined a fantasy league run by some friends from a sports mailing list.†

---

* At the risk of alienating some readers, my primary allegiances are to the New York Giants in football and the Syracuse Orange and Maryland Terrapins in college basketball (I grew up in New York, south of Syracuse, and went to graduate school at Maryland). I don't follow baseball all that closely, but to the limited extent that I pay attention to it, I vaguely root for the Yankees.

† Making me one of the few people in America not employed by ESPN who can honestly say he took up fantasy football for work.

I maintained a few extra rules for myself—I refuse to have any players from the Dallas Cowboys or Philadelphia Eagles on my team—but I figured that the best way to get a sense of the phenomenon was from the inside. On the whole, it has been fun, particularly in the way that it gives meaning to games that I otherwise wouldn't care about. Tampa Bay's playing Oakland in football would ordinarily hold no interest for me, but in 2012, I had Tampa's running back Doug Martin on my fantasy team; his great game against the Raiders was a highlight of the season. Of course, there are also some hard reality checks, like getting thoroughly trounced—twice!—by the ten-year-old son of one of the other fantasy players.

The most striking thing about dabbling in fantasy football has been the way it more than confirmed my initial idea about the role of statistics and the degree of mathematical sophistication required. Deciding who to draft and who to play from week to week turned out to be even more complicated than I had expected; as I don't have a great deal of free time, I'm not the most hands-on owner, but I do make some effort to swap players going up against a particularly tough defense out in favor of those playing notably inept defenses, and the like. And trying to understand the personnel moves made by some of the more active guys in the league made clear just how complicated the game can be if you're willing to put in the time.

Running a fantasy football team has also given me a greater appreciation of statistics in sports. It might seem surprising, given my background in a highly mathematical science, but I've never been all that fond of sports statistics, both because it seemed a little too much like work and because of the inherent uncertainties (which we'll discuss more in the next chapter). Running a fantasy team and reading statistically informed writers like Bill Barnwell at *Grantland*, however, has convinced me of the value of at least some of the number-crunching that goes on around modern sports.

So, to return to Carl Sagan's quote about the relationship between sports fans and statistics, as dry and dusty as the box scores cluttering up the sports pages may look at first, once you begin to look into them, they are actually a vivid and effective way of communicating information. The millions of people letting their inner scientists help them pretend to run sports teams in their spare time are a testament both to the effectiveness of statistics as a communication medium and to the ability of properly motivated fans to grasp the mathematical details.

Chapter 14

DAMNED LIES AND STATISTICS

Figures often beguile me, particularly when I have the arranging of them myself; in which case the remark attributed to Disraeli would often apply with justice and force: "There are three kinds of lies: lies, damned lies, and statistics."

—Mark Twain, *Chapters from My Autobiography*

While many people voluntarily seek out and read statistical analysis of sporting events, most other forms of statistics are viewed with distaste. They're nearly impossible to escape, though, as virtually any public policy decision can only be understood with statistical methods. The effects of a decision on any given individual are all but impossible to predict, but the aggregate effect over the entire population of a city, state, or country can often be accurately predicted using statistics. Accordingly, national and international news media are full of statistics of one sort or another. The overall effect can be bewildering to those not comfortable with the underlying math.

Statistics also play a role in public confusion about science, particularly in the reporting of health stories. Even the most casual consumer of news has undoubtedly noticed a sequence of seemingly contradictory findings—doctors say that alcohol is bad for you, except maybe red wine is good for you, but rats given beer to drink live longer; obesity is a national health crisis blamed for all sorts of problems, except people who are slightly overweight live longer; this nutrient or that one prevents cancer and heart disease, except no, it doesn't. This confusion stems from the statistical nature of medical studies and sometimes even trips up scientists, as we'll see later.

Given the importance of statistics for public policy and their ability to sometimes confuse professional scientists, it's no surprise that statistics have a vile reputation, as demonstrated by the famous taxonomy of untruth attributed to the Victorian-era prime minister of the United Kingdom, Benjamin Disraeli.* And statistics are undeniably used by canny and cynical politicians and corporations to create false impressions in the public. Confronted with scientific evidence favoring a particular policy, opponents will trot out statistics of their own and point to the history (real or imagined) of public reversals and contradictory advice to create uncertainty about what science actually says.

As difficult as statistics may be to deal with, however, they have their fans, in greater numbers than you might think. One of the biggest winners of the 2012 election cycle in the United States (aside from those elected to office, of course) was a consultant-turned-poker-player-turned-baseball-statistician-turned-political-analyst named Nate Silver. His blog, *FiveThirtyEight*, was picked up by the *New York Times* and read by millions—at the peak of the election season, something like

---

* Like many great lines attributed to famous people, there's little evidence that Disraeli ever actually said this—the attribution to him appears to have originated with the Mark Twain quote at the start of this chapter—but it's too good a phrase to resist.

20 percent of the daily visitors to the *Times* website read Silver's blog.[†]

Surprisingly, what drew people to *FiveThirtyEight* was not tawdry tabloid journalism or muckraking investigative reporting, but hard-core math. Silver's claim to fame before *FiveThirtyEight* was as the inventor of a statistical evaluation system for projecting the effectiveness of Major League Baseball players. *FiveThirtyEight* grew out of his decision to apply the same methods to forecasting elections. Tracking dozens of polls that asked voters about the presidential and congressional elections, he combined the results with some economic and demographic data to predict the outcomes of those elections. He wasn't the only one doing this—Sam Wang of the Princeton Election Consortium and Drew Linzer of Votamatic ran similar websites with statistical projections—but Silver became the public face of the political modeling movement.

Silver's foray into political analysis drew criticism from established political commentators, several of whom scoffed loudly and publicly at the idea that mere number-crunching could replace years of journalistic experience and inside sources. Several pundits—Peggy Noonan, David Brooks, and others—derided Silver's prediction of a decisive Obama victory and offered their own take from insider interviews and "gut feelings."[‡] The modelers had the last laugh, however, as

---

[†] See Peter Kafka, "Here's What the New York Times' Nate Silver Traffic Boom Looks Like," *All Things Digital*, November 6, 2012, http:// allthingsd.com/20121106/heres-what-the-new-york-times-nate-silver -traffic-boom-looks-like/.

[‡] Noonan predicted a Romney victory on the basis of, among other things, her interpretation of Obama's facial expression and the number of Romney signs she saw in people's yards (Brett Logiurato, "Peggy Noonan Predicts a Romney Victory in the Most Anti-Nate Silver Column Imaginable,"

Silver's model correctly predicted the outcome of the presidential election in all fifty states, and thirty-one of the thirty-three Senate races. Wang did even better, correctly calling all the races Silver did, plus the Senate seat in North Dakota.

The huge readership for Silver's blog and his best-selling book about statistical predictions, *The Signal and the Noise*, demonstrates that despite the bad reputation of statistics in public policy since the days of Twain and Disraeli, there is a large audience for statistical analysis of elections.* Making complicated numerical models to predict elections is not likely to become a popular pastime, but nonscientists can absolutely read, follow, and even enjoy discussions of the underlying issues and become better informed citizens as a result.

In this chapter, we'll look at some simple examples of statistics in action, using the election of 2012 as a starting point. We'll also see that the techniques used to project elections are closely related to common techniques in scientific analysis and communication. A basic understanding of these tools can help you avoid being lied to by politicians and pundits and can also lead to a better ability to understand and evaluate scientific results.

THE SCIENCE OF UNCERTAINTY

One of the most prominent dismissals of Nate Silver and other statistically based election forecasters came from MSNBC host

---

Business Insider, November 5, 2012, www.businessinsider.com/peggy -noonan-romney-obama-prediction-electoral-college-map-wall-street -journal-2012-11). Brooks wrote that "when [pollsters] start projecting, they're getting into silly land" (Dylan Byers, "Nate Silver: One-Term Celebrity?" *Politico*, October 29, 2012, www.politico.com/blogs/media /2012/10/nate-silver-romney-clearly-could-still-win-147618.html).
* *The Signal and the Noise: Why So Many Predictions Fail—But Some Don't* (New York: Penguin, 2012).

Joe Scarborough, whose on-air rant became emblematic of the conflict between statisticians and pundits:

> Nate Silver says this is a 73.6 percent chance that the president's going to win. Nobody in that campaign thinks they have a 73.6 percent—they think they have a 50.1 percent chance of winning.
>
> . . . Anybody that thinks that this race is anything but a tossup right now is such an ideologue [that] they should be kept away from typewriters, computers, laptops, and microphones for the next ten days, because they're jokes.[†]

One fairly charitable interpretation of this statement is that Scarborough confused two different kinds of percentages: mistaking the probability Silver's model gave for an Obama win for a prediction of the share of the vote Obama would win.[‡] Certainly, nobody in the campaign thought that a candidate would win 74 percent of the popular vote—but then Silver didn't think this, either. What he predicted was a 74 percent chance that Obama would win the election with a bit more than 50 percent of the popular vote. Scarborough mistook a statement about the *uncertainty* of a prediction for the prediction itself.

[†] Quoted many places, including Christian Heinze, "Scarborough Fires Shot at Nate Silver's Model," *The Hill*, October 29, 2012, http://gop12.thehill.com/2012/10/scarborough-fires-shot-at-nate-silvers.html.

[‡] Less charitable interpretations abounded, with *The Atlantic Wire*'s headline for its post about Scarborough's tirade being a good example: Elspeth Reeve, "People Who Can't Do Math Are So Mad at Nate Silver," *Atlantic Wire*, October 29, 2012, www.theatlanticwire.com/politics/2012/10/people-who-cant-do-math-are-so-mad-nate-silver/58460/. On the more charitable side, there is a philosophical issue regarding the validity of assigning probabilities to singular events like elections, but given the specific numbers he used, I don't think that's what Scarborough was getting at.

The fundamental problem of statistics and the abuse thereof is uncertainty. If we could track with perfect certainty everything that goes on in a political campaign or scientific experiment, there would be no need for statistical methods—we'd just say, "This is exactly what happened," including all of the relevant factors, and that would be the end. Even relatively straightforward scientific experiments, however, are subject to uncontrollable variations (as discussed in Chapter 9), and elections are the result of individual decisions by millions of people. Our inability to keep track of these things introduces an inescapable element of randomness to the outcome. Statistics are just mathematical tools for dealing with that randomness and, just as importantly, for quantifying the uncertainty in predictions of random events.

As it turns out, most people don't deal well with randomness. Misconceptions about randomness are the foundation of sports punditry and the enormous profits raked in by casinos. Fortunately, though, it's relatively easy to investigate simple random systems directly: One of the canonical examples of randomness is the flipping of a fair coin. You can easily demonstrate the basic properties of randomness and uncertainty with pocket change.

So: find a coin, and flip it 25 times, recording the number of times it comes up heads. For a fair coin, that should be a fifty-fifty proposition, so you would expect it to land heads up about half of the time. But of course, for an odd number of flips, it can't be heads *exactly* half of the time—this would require the coin to show heads 12.5 times, which is impossible. The closest you can come is either 12 heads and 13 tails, or 13 heads and 12 tails. If you just did the experiment, though, the most likely result is something other than a 13–12 split.

In fact, if you repeat the 25-coin-flip experiment multiple times, you'll get a 13–12 (or 12–13) split a bit less than a third of the time.* Those are the two most likely individual results, but the real distribution encompasses a much wider range. A truly fair coin will give uneven results far more often than most people expect, and the same is true of all random processes.

This does not mean, however, that absolutely anything is possible. The range of possible outcomes is tightly constrained, and we can use statistical tools to tell us precisely how much variation to expect. For the experiment of flipping a coin 25 times, we can say that the total number of heads will be between 10 and 15 just about 70 percent of the time. This prediction is easily verified with a coin or a random number generator.

The size of that range increases as the number of flips increases, but only as the square root of the number of flips—so if we were to quadruple the number of flips, we would only double the size of the uncertainty range. If you flip a fair coin 100 times, you will get between 45 and 55 heads around 70 percent of the time. (This gets a little tedious, but is easy to verify.) These kinds of predictions are the fundamental elements of statistics: We cannot predict the exact outcome of a particular set of coin flips, but we can make very confident predictions about ranges of possible outcomes.

The percentages Nate Silver used and Joe Scarborough misunderstood are statements about uncertainty in the range of possible outcomes according to Silver's model. The claim was that there was a 74 percent chance that the final election results would fall within a range that would give Obama a victory. While the detailed modeling process is more complicated for elections than coins, the underlying idea is the same: Statistical analysis allows us to make confident statements about the range of possible results of a random process. Anybody who can do the math can make the same sort of confident

---

* The exact probability is about 31 percent.

predictions; thus, Sam Wang offered to eat a bug on camera if Mitt Romney won either Minnesota or Pennsylvania, and a "really big bug" if Romney won Ohio.* His statistical predictions gave not only an Obama win as the most likely outcome in those states, but also an uncertainty range that made insect-eating extremely unlikely.

POLLING THE DICE

It may seem unusual to talk about something like an election as random—after all, people go to the polls knowing who they plan to vote for.† But while something like voting may be completely deterministic on an individual level, at the level of the entire nation or even a single precinct, the result is effectively random in terms of prediction. This is why pollsters have to use statistics and the key to how Silver and Wang were able to generate their probability estimates.

The randomness of election predictions comes partly from genuine chance—people who leave the house intending to vote, but don't make it to their polling place because an emergency came up at work, say—but mostly due to polling errors. One major source of trouble is recency: A poll is necessarily a snapshot of opinions at the time of the poll, and these may change in the time between the poll and the election. An even more important issue is sampling: A preelection poll can't ask *every* voter who he or she plans to vote for—not only would that be prohibitively expensive, but it would remove the need to have the election in the first place. Instead, pollsters select a

---

* Sam Wang, "How Likely Is a Popular-Vote/Electoral-Vote Mismatch?" *Princeton Election Consortium,* November 3, 2012, http://election.princeton .edu/2012/11/03/how-likely-is-a-popular-voteelectoral-vote-mismatch/.

† Well, except for a tiny number of possibly mythical "swing voters" who can't seem to make a commitment.

sample of voters and extrapolate from the responses to predict the final vote. This process is necessarily uncertain, though, and depends strongly on who the pollsters happened to select for their random sample.

There are well-defined rules for determining the uncertainty introduced by extrapolating from a small random sample, but the simplest rule of thumb is very similar to the uncertainty in the coin-flip results just discussed. The inherent sampling uncertainty in a poll is one over the square root of the number of people polled. A poll that asks 100 people who they plan to vote for will give an uncertainty of 0.1, or ±10 percent, while a poll that asks 10,000 people who they plan to vote for will give an uncertainty of 0.01, or ±1 percent.[‡]

The polls reported in national media generally survey around 1,000 voters, giving an inherent uncertainty of about ±4 percent. That number is a balance between economics (it's cheaper to survey a small number of people), the desire for better results (more respondents means a smaller sampling error), and the influence of other sources of error. The simple formula is only completely accurate for a truly random and representative sample of the whole population, which is a very difficult thing to do. There are countless ways for the sampling to go wrong, and much of the work of professional pollsters comes in not only trying to get the most random and representative sample possible, but also trying to correct for the ways the sample might deviate from pure randomness.

[‡] This slow decrease in uncertainty is one good reason to be skeptical of some arguments about the sports statistics discussed in Chapter 13. Even for baseball, with the largest sample size of any major American sport, the intrinsic uncertainty in a player's batting average for an entire season is around 0.020. Arguments that turn on smaller differences than this (e.g., most of the batting title awards in history) are not terribly meaningful for comparing the ability of players.

The institutional practices of modern political polling can be traced to two epic disasters related to these errors. The first came in 1936, when the *Literary Digest* predicted Republican Alf Landon to defeat incumbent president Franklin Roosevelt in a landslide. The prediction was based on a poll in which several million Americans were asked to return cards indicating their preference. That same year, an independent pollster named George Gallup predicted a Roosevelt landslide on the basis of polling only a few thousand voters. Gallup came in for a bit of mockery at the time, but like the statisticians of 2012, he had the last laugh—Roosevelt won, as predicted, by one of the biggest margins in electoral history.

Why was Gallup successful when the *Literary Digest* failed? The *Digest*'s huge sample suffered from two main problems: First, it was drawn from telephone directories, club memberships, and magazine subscriptions; these subsets of the population may have been biased toward more affluent voters who were likely to vote Republican. Second and equally important, the final sample of over two million responses came from an initial poll of more than ten million, a response rate of less than one in four. Such a low response rate can introduce a *nonresponse bias*: Although the original sample may have been reasonably random, the people who actually responded presumably shared some common characteristics that led them to respond and, as a result, were not a truly representative sample. Gallup had a smaller sample, but his was ultimately more representative of the general electorate, and thus he got a more accurate prediction.

Still, a dozen years later, in 1948, Gallup wound up on the wrong side of another electoral debacle. One of the most iconic images in American political history is the photo of a jubilant Harry Truman holding aloft a copy of the *Chicago Tribune* with the headline "Dewey Defeats Truman." Truman had, in fact, won the election over his opponent, Thomas Dewey, a few days earlier.

The erroneous headline came about because of several factors: Thanks to a labor dispute, the *Tribune* needed to go to press a few hours earlier than usual and thus had to pick a headline before the complete results were in. It opted to predict a Dewey victory in light of the gut feeling of its Washington correspondent Arthur Sears Henning, who had correctly predicted elections in the past. Henning's confidence in Dewey was bolstered by Gallup polls showing Dewey with a commanding lead. Dewey's lead was so consistent, in fact, that Gallup had stopped doing new polls some weeks earlier, in the belief that the results were set and unlikely to change.

These two great failures of prediction helped set the template for modern American election polling: To avoid Gallup's mistake of 1948, polling firms now conduct polls right up until the day of the election. And to avoid the *Literary Digest*'s errors, they use frequent polls of smaller, easier-to-control samples—almost every day during the 2012 election saw the release of several new polls based on samples of a few thousand voters.

Of course, if polling were simple, it wouldn't be a business. Turning these few thousand responses into a prediction of millions of eventual votes involves a host of proprietary adjustments and corrections. It's not as simple as counting up the number of people professing an intent to vote for Obama and dividing by the total size of the sample. Instead, most polling firms calculate a weighted average of the results (as discussed in Chapter 13), to account for party identification (e.g., if their random sampling catches too many registered Democrats, the polls will give greater weight to the responses of Republicans to balance things out), gender, race, age, location, and voter registration status. The exact methods, for the most part closely guarded trade secrets, provided a lot of material for Silver's election blogging, in which he discussed "house effects" and the like. Poll aggregators like Silver, Wang, and Linzer use a similar process at a higher level, calculating a weighted average of all the available public polls, with the weights and other adjustments determined by the historical performance of each polling firm and the date of the poll (giving greater weight to more recent polls).

ELECTORAL D&D:
POLLS AND MONTE CARLO SIMULATIONS

For predicting the 2012 election, another complication was introduced by the US Constitution. As every grade school student learns, and anybody who lived through the election of 2000 will never forget, the president is not determined by the popular vote, but is elected by the US Electoral College.* The popular vote determines the winner of each state, with the state winners accruing electoral votes according to the congressional representation for that state. This arrangement complicates the business of predicting elections—it's not as simple as averaging polls and assigning uncertainty according to sample sizes.

The confidence numbers that Silver reported drew on another technique common in science: computer simulations. In addition to statistical analysis of the polls themselves, Silver ran and reported numerous simulations of the race, using the state-by-state polling information to set the parameters. This method, known as a *Monte Carlo simulation*, shows up in almost every field of science when scientists are dealing with situations where direct calculations are too complex.

---

* Believe me, given the number of "hanging-chad" jokes I had to endure in late 2000 and early 2001, I would really love to forget that particular debacle.

The exact details of the simulations are not public, but the essential idea is to use the aggregated state-by-state polls to determine the most likely vote share for each candidate and an uncertainty in that share. A single run of the simulation then uses a computer to randomly select an exact value from within the uncertainty range for each state. The "winner" of each state in the simulation gets that state's electoral votes, leading to a total simulated electoral vote. These simulated elections are repeated tens of thousands of times, and the results of all the simulations are compiled to produce the final prediction: The 74 percent probability of an Obama victory means that 74 percent of those thousands of simulations came up with Obama winning.[†]

This scheme may seem like a bit of a cheat, as if you're calling an election on the basis of the rolling of thousands of dice. In reality, though, the method is extremely robust and sees heavy use in other areas of science, particularly in subfields of physics and chemistry where complex interactions of many factors make simple mathematical formulas impossible. Particle and nuclear physicists, in particular, make heavy use of Monte Carlo simulations for understanding their detectors.[‡] The LHC collaborations discussed earlier in the book construct detailed models of their detectors, including all the many layers and components. These models generate simulated data tracks for "collisions" that produce well-understood particles using random numbers generated for each stage. As the simulated particle enters each simulated component, the computer decides whether it is recorded, using detection probabilities based on the measured performance of the

---

[†] The exact number of simulations was sometimes kind of strange: The number 25,001 showed up several times. Apparently, that last "1" makes all the difference.

[‡] The first Monte Carlo methods were developed in the 1950s and were used in nuclear weapons development.

components. These simulated data tracks—the total amount of simulated data is comparable to the data collected from actual collisions—are used to understand the background from Standard Model particles against which new particles might appear. The models help the scientists estimate how often they expect a particular set of collision products in their detector, so when they see those products show up a few hundred extra times in collisions at a particular energy, those collisions can serve as evidence of the Higgs boson.

THE SIGNIFICANCE OF STATISTICAL SIGNIFICANCE

Statistical statements about uncertainty ranges play a crucial role in science. Scientific measurements are inherently subject to random fluctuations, and thus, strictly speaking, all scientific results are statistical in nature. There is always some chance that a particular measurement doesn't really reflect what the scientists thought they were measuring, but instead reflects some chance fluctuation in the environment—a polling sample that appeared random but had a hidden bias or a physics measurement perturbed by a passing train (Chapter 9). To deal with this uncertainty, scientists define the range of possible outcomes predicted by a model, and calculate the probability that a given measurement will fall within that range ("when you flip a fair coin 100 times, there is a 70 percent chance that between 45 and 55 of the tosses will come up heads"). Given a set of repeated measurements, then, we can say whether the actual results are consistent with the prediction at a level that makes us trust that model.

Of course, even for people comfortable with math, constantly dealing with ranges of numbers gets tiresome, so most fields of science have some standard for defining a threshold separating "real" results from mere random noise. This standard is typically defined in terms of the probability that the world is actually well described by previously known science (the *null hypothesis*), but this particular experiment just happened to get (un)lucky enough to end up with a measurement that makes it look like some new phenomenon is at work. In other words, what is the probability that this result was just a random fluctuation? If that probability is low enough, the result is said to be *statistically significant*.

The exact threshold for statistical significance is a matter of local convention. In many life-science fields, the standard is 5 percent (about 60 out of 100 coin flips coming up heads would suggest the coin is biased toward heads at this level of significance), while in particle physics, the standard is *five sigma*, or about 1 chance in 3.5 million (around 75 out of 100 coin flips coming up heads). The difference in standards reflects the difficulty of performing experiments. Although this difference between life sciences and physics might seem backward, given the cost of building a giant particle accelerator, once you have the LHC, you can easily record trillions of collisions, while it takes an extraordinary effort to run a new medical study with ten thousand participants. It's much easier for particle physicists to do enough experiments to see exceedingly unlikely events, so they need a more stringent standard.

The conventions of statistical significance are responsible for many common misunderstandings regarding scientific results reported in the media. One problem is simply linguistic confusion—in everyday speech, *significant* carries a connotation of "large," so when scientists say that a result is statistically significant, many people will assume that means the effect is dramatic. In fact, statistical significance and the size of the effect are largely independent of each other. A well-designed experiment can produce statistically significant results regarding extremely small effects—the Higgs boson was detected with a very high level of statistical significance, but only showed up hundreds of times out of trillions of collisions at the LHC. And particularly difficult or poorly designed experiments can fail to reach statistical significance even when effects are fairly large—a poll of only one hundred voters, for example, would need the difference between candidates to be around 20 percent to claim statistical significance by the common standards in social science, but even smaller vote margins than that are regarded as landslide victories.

The second source of confusion has to do with methodology and sometimes trips up scientists as well as reporters. If the standard for claiming to detect a new effect is a 5 percent probability that previously known science would produce your data by chance, then you would expect roughly one study in twenty to produce "significant" results purely by chance. If you look for an effect in one hundred different trials, then, some of them are bound to look like they succeeded, even if the effect in question doesn't exist at all.

Statistical uncertainty partly accounts for the apparent yo-yoing of medical science, as there are huge numbers of experiments under way at any given time. The problem is further compounded by the far greater difficulty in publishing negative results—showing that a new effect *does not* exist— than in publishing positive ones. Even research that reports a failure to replicate a previously published result can face an uphill struggle to get published. This combination of random chance and the bias for positive results is the origin of those strings of seemingly contradictory results, reporting effects that first go one way and then go the other. False-positive results get published and attract a lot of publicity, whereas failures to replicate these results receive little attention, until another dramatic result going in the opposite direction (which may itself be a fluke) surfaces.

Responsible scientists try to avoid this problem by using multiple lines of evidence that point in the same direction before claiming the discovery of a new effect. Even so, spurious "significant" results are bound to turn up—even in particle physics, where there have been retractions of five-sigma results claiming new particles.

It's a good idea, then, to follow two rules regarding splashy new results reported in the media: First, look at both the size of the effect and the statistical significance, because there is a complex interplay between them. A result can be statistically significant without being practically significant—a small increase in the probability of getting an exceedingly rare disease may pass the statistical threshold to be reported as a real effect without making enough difference in your risk of death to be worth changing your behavior.* On the other hand, an effect may appear to be large, but if it is only marginally significant, you would do well to wait for more data.

The second important rule to remember is to always regard single results as provisional. One new report may be interesting, but it's not scientifically solid until it has been repeated multiple times in independent tests. Repeatability can be somewhat difficult to discover—"Scientists Confirm Earlier Result from Other Scientists" is not a headline you'll see a lot—but in most cases, it's worth a little extra research before making major lifestyle changes.

---

* On the other hand, given a large enough population, even a small effect may have expensive consequences for public health outcomes and may demand policy changes if the effect is statistically significant.

By these standards, then, how do electoral statisticians like Nate Silver, Sam Wang, and Drew Linzer stack up as scientists? Obviously, they did much better than Joe Scarborough, Peggy Noonan, and David Brooks, but even by higher standards, they're doing reasonably well. All three statistical models successfully predicted the outcomes of a wide range of state and national elections in 2012, and Wang and Silver also performed well in 2008. The poll modelers have managed at least some degree of independent replication, then, although a couple more election cycles would be a better test.

The poll modelers also show the importance of objectivity and quantitative measures to the process. Wang, Linzer, and Silver used a collection of all the publicly available polls, combining them according to simple but strict mathematical rules. This method averages out what may be fluke results from individual polls or systematic biases from the internal methods used to adjust the samples. The professional pundits, on the other hand, were more likely to seize on single poll results that happened to agree with their personal preferences or even to discard polling altogether in favor of fuzzy qualitative indicators like yard signs and facial expressions. Although these methods may make for more dramatic television and higher ratings, the techniques are highly susceptible to personal bias.

More importantly, though, the statisticians' engaging approach to political analysis shows that statistics don't need to be arcane and incomprehensible. Figures may have beguiled Twain (note, however, that he was a pretty sharp guy fond of exaggeration for humorous effect). But if you pay attention to your inner scientist, you can avoid being lied to with statistics. In fact, detailed analysis of statistics can be just as entertaining as any other form of commentary, while being far more informative and reliable.

SCIENCE IS NEVER OVER

For most of this book, we've talked about the many things that science is and how scientific processes are at the heart of a wide range of activities that may not at first seem scientific. I hope that through these stories, I've shown you some of the many ways in which you already use your inner scientist and some additional ways your inner scientist might be useful.

There are, however, a lot of persistent myths and misconceptions about how science works, who can do science, and generally what science is. Some of these we have already seen explicitly, while others have been implicit in the stories we've looked at. So it's worth devoting a little time to talk about things that science is *not*.

SCIENCE IS NOT EXCLUSIVELY WESTERN

The explosive growth of modern science is often traced to the Renaissance in Europe, and the institutional structure of modern science owes a great deal to the European system of universities that began in the late medieval period. Most of the world's leading scientific institutions are currently located in Western Europe and the United States. These facts have all too often led people to declare that science is somehow essentially Western.

This claim takes many forms, some fairly harmless, like abstract academic arguments about the social construction of science, and others less so, as when whole countries reject science or scientific findings for reasons of national pride. The Stalin-era adoption of the discredited biological ideas of Trofim Lysenko set biology in the Soviet Union back for decades and contributed to famines in the USSR and China. The rejection of the idea that HIV causes AIDS by the Thabo Mbeki administration in South Africa delayed the implementation of effective prevention programs in that country, and some studies claim this misinformed policy led to hundreds of thousands of preventable deaths.

A particularly unfortunate version of Western bias is the idea that the European civilization is innately and uniquely suited to the practice of science—an attitude that leads to ugly statements from prominent people who ought to know better. Nobel laureate James Watson was forced to resign as director of the Cold Spring Harbor Laboratory in 2007 after his disparaging comments about people of African descent provoked outrage. And prominent biologist and outspoken atheist Richard Dawkins sparked a good deal of controversy in 2013 for a dismissive comment on Twitter comparing the number of Nobel prizes won by Muslims with the number won by scientists at Trinity College in the United Kingdom.

The idea that science is inherently Western is, of course, complete nonsense. We have clear evidence of humans doing science going back thousands of years before Europe— let alone Western culture—was a recognizable concept. The Neolithic passage tomb at Newgrange in Ireland, built around 3000 BC, is a mound of roughly two hundred thousand tons of rock containing a passage that is precisely aligned with the rising Sun on the winter solstice, the shortest day of the year. At dawn on the solstice, a beam of light passes through a notch above the door and illuminates the central chamber of the tomb for a few minutes, the only light the chamber receives all year. No one knows just who Newgrange's builders were or exactly what they were doing, but their legacy remains in their stone monuments and speaks clearly of ancient science. They made careful measurements of the motion of the Sun across the sky over a period of many years, probably centuries, passing that knowledge along through generations. They devised models of that motion that predicted the position of the rising Sun at significant future dates and refined those observations and models to such a degree that their monuments still work thousands of years after their civilization vanished.*

Every culture we have any knowledge of has practiced astronomy, making careful observations of the motion of heavenly bodies, making models to predict those motions, testing and refining those models, and passing them down through generations. From ancient Greece and China to Mesoamerican civilizations like the Maya and Inca, and even the Polynesians in the Pacific, who used a complicated system of celestial navigation to sail between islands hundreds of miles apart, great civilizations have always had a sophisticated understanding of astronomy.

In fact, the European dominance of modern science would not have been possible without the Muslims Dawkins disparaged. Many of the great works of Greek and Roman science

---

* Similar astronomical alignments are found at other sites from roughly the same era, like another passage tomb at Maeshowe in the Orkney Islands and, most famously, the circle of standing stones at Stonehenge.

would have been lost in the chaos following the collapse of the Roman empire had they not been copied and preserved by Muslim scientists. The rediscovery of many of these manuscripts, translated back into European languages from Arabic, played a significant role in the development of European science, and Muslim scientists made great advances in astronomy, optics, medicine, and mathematics during the medieval period.* Much of the vocabulary used for science and math— terms like *algebra, algorithm, alchemy,* and *azimuth* and a long list of names for stars—comes from Arabic roots, and the modern symbols for writing numbers were developed in India and brought to Europe by Muslim Arabs.

Groundbreaking scientific discoveries have been made by people from many cultures, even in the modern era. One of the most important developments in the history of nuclear and particle physics was the meson theory, for which Japanese physicist Hideki Yukawa won the 1949 Nobel Prize in Physics. Yukawa's theory explained the interaction between protons and neutrons in the nucleus as arising through the exchange of massive particles dubbed *mesons.* His theory set the stage for the later work of his countryman (and Kyoto University classmate) Sin-Itiro Tomonaga on QED (Chapter 11) and pretty much all subsequent subatomic physics.

In astrophysics, one of the key discoveries in our understanding of stellar evolution came out of work done by a young Indian physicist, Subrahmanyan Chandrasekhar, on a steamship voyage to England in 1930. Chandrasekhar spent much of the voyage in his cabin, working out the physics of electrons in white dwarf stars, including the effects of relativity; this work

---

* To be fair, Dawkins's offensive Twitter comments included a sarcastic reference to these contributions.

eventually led to the *Chandrasekhar limit*, the maximum mass a white dwarf star can have before collapsing into a neutron star (like the pulsars discussed in Chapter 4). Chandrasekhar won the Nobel Prize in 1983 for this work, among many other important contributions to astrophysics. Chandrasekhar's uncle Chandrasekhara Venkata Raman also won a Nobel, in 1930, for seminal research into the interaction between light and matter. A third Indian physicist of that era, Satyendra Nath Bose, was the first to work out the mathematical properties of a class of quantum-mechanical particles now dubbed *bosons* in his honor. These particles include photons, the recently discovered Higgs boson, and many types of atoms. Bose's work is essential for understanding the operation of lasers, superconductors, and much of particle physics.

Science is not a Western creation. Science is a universal human activity, practiced and advanced by people from many cultures—we all have an inner scientist, regardless of where we were born or how we were raised. As we've seen throughout this book, it's almost impossible to get through a day without making some use of the scientific process. The current Western domination of institutional science is little more than a historical accident, and the continuing rise of science in China and India, among other countries, may soon relegate Western dominance to history entirely.

SCIENCE IS NOT JUST FOR BOYS

Another persistent myth about science is that it's an intrinsically male activity. Again, this is a depressingly common source of stupid statements from powerful people. In 2006, the president of Harvard, Larry Summers, resigned after a five-year tenure full of conflicts with his faculty. One of the most public of these was the controversy that exploded after a 2005 speech in which he suggested that women were less likely than men to have the necessary ability to succeed at the highest levels of science. Again, this idea is nonsense—as we've seen in this book, many important scientific discoveries were made by women.

There are undoubtedly more men than women in academic science, but this is mostly the result of a long history of active opposition to women in academic science. Emmy Noether, one of the greatest mathematicians of the twentieth century, spent some of the most productive years of her career in unpaid positions. She completed her dissertation on mathematics at the University of Erlangen in 1907, then spent several years working there as an unpaid substitute lecturer for her father and other professors in the department.* During this time, she pursued mathematical research on her own and made some significant contributions building on the work of David Hilbert, one of the most eminent mathematicians of that era.

Noether's work caught Hilbert's attention, and in 1915 he and Felix Klein invited her to come work at the University of Göttingen. Despite Hilbert's backing, the faculty refused to grant her a paid position, asking, "What will our soldiers think when they return to the university and find that they are required to learn at the feet of a woman?" In exasperation, Hilbert retorted, "We are a university, not a bath house." The only concession he was able to wring out of the administration, though, was permission for Noether to teach without pay. Hilbert was listed as the instructor for her courses, with Noether

---

* Her dissertation was finished not long after rules barring women from receiving higher degrees were relaxed. In the earlier part of Noether's career, she was not officially allowed to take classes for credit, but was permitted only to audit them.

as his "assistant." She wasn't formally approved to teach students on her own until 1919 and didn't receive a salary from the university until 1922, when she was granted adjunct status.[†]

Despite these obstacles, Noether's work during these years had a revolutionary impact, particularly in the field of abstract algebra. The work gained her international recognition, culminating in a plenary address to the International Congress of Mathematicians in 1932. Over her years in Göttingen, she attracted a devoted following of students, and the list of "Noether's boys" reads like a who's who of abstract algebra.

Noether is also held in high regard within physics, where she's best known for a theorem she proved in 1915 showing a deep relationship between symmetries and conservation laws in physics. Physicists constantly rely on the idea of conservation of energy—that the total energy of a system of interacting objects will always add up to the same value. Noether showed that this rule is a consequence of the fact that the laws of physics do not change in time. Similar relationships hold for other symmetries—conservation of momentum comes about because the laws of physics do not change as you move through space, and conservation of angular momentum holds because the laws of physics do not change as you turn in different directions. In fact, any time there is a symmetry in the laws of physics—something you can change without affecting the underlying rules—there will be a conserved quantity. As a result, Noether's insight has completely changed the way theoretical physicists look at the world.

[†] She never received full faculty status in Germany. She only officially became a professor after the Nazis took power in 1933 and banned Jews from holding academic positions, even low-status ones. At that point, the Rockefeller Foundation arranged a full professorship for her at Bryn Mawr College in Pennsylvania.

Another woman who made revolutionary contributions to our understanding of symmetry in physics is Chien-Shiung Wu, this time from the experimental side. A product of the first generation of girls allowed to receive formal education in China—Wu's father founded a school for girls specifically to be able to teach her—Wu distinguished herself as one of the country's finest undergraduate physics students in the 1930s. As China did not have Ph.D. programs in physics at that time, she came to the United States in 1936 and joined Ernest Lawrence's group at Berkeley, one of the world's best laboratories for experimental particle and nuclear physics. Even in that elite company, she distinguished herself. During World War II, Wu was the only Chinese scientist invited to work on the Manhattan Project; one version of the story has it that Enrico Fermi was struggling with a thorny problem in the uranium enrichment process and was told by another physicist that if he wanted to sort this out, he should "ask Miss Wu." Wu found the problem, and her contributions were essential to the success of the atomic bomb program.

Wu's greatest contribution, however, came after the war, when she joined the faculty at Columbia University. In 1956, her Columbia colleague Tsung-Dao Lee and Chen-Ning Yang at the Institute for Advanced Study began looking into the idea that *parity symmetry*, then believed to be a fundamental principle of physics, might occasionally be violated. Lee and Yang realized that contrary to conventional wisdom, some beta decay reactions ought to show a preferred direction—that when the particles involved fell apart, they would be more likely to spit out an electron in one particular direction than the other, relative to the magnetic poles of the particle. This effect would be subtle and would require exacting care to be detected experimentally, so Lee and Yang approached Wu about looking for it.

Wu dropped everything—including a long-planned trip to China to visit her family there—and she and her Columbia students and colleagues at the National Bureau of Standards spent six months designing and assembling an experiment to measure this violation of parity symmetry. By Christmas of 1956, they had their experiment ready to go and, in early January, they announced the result: Their experiment showed unequivocally that decaying nuclei of cobalt did, indeed, show a preferred direction. A less skilled experimenter than Wu might not have detected the result or might not have been believed, but her talent and reputation were up to the task. The discovery of parity violation sparked another great revision of our understanding of fundamental symmetry in physics, and the idea is now central to explaining why the universe contains more matter than antimatter.

The stories of Noether and Wu, and all the many other great discoveries made by women (often in the face of great professional and personal adversity), demonstrate clearly that women can and do excel at science. Again, this conclusion should not come as a surprise, because the core of science is a simple and universal mental process shared by all humans. None of the activities we've looked at in this book are intrinsically male, and all of them make use of scientific thinking.

The perception that science is a male activity is sadly persistent, though, mainly because of the insidious effects of socialization. Adults who think of science as a male profession pass that message on to children, sometimes consciously, but often almost subliminally. This bias is particularly visible in toy stores and bookstores, where science-related toys are almost invariably packaged and marketed in ways that appeal to boys, while girls are pushed toward dolls and other domestic toys.

The gendered marketing of science toys is not a result of any inherent difference in interests. My daughter's preschool class produced an end-of-the-year composite photo of each child holding a sign saying what he or she planned to be as a grown-up, and I was enormously pleased to see that hers read "scientist." And when she has other girls from her school over for play dates, one of their favorite activities is playing with her science set, "inventing new kinds of water" by mixing various items in test tubes. Science is just as fascinating for girls as boys, no matter what toy makers may think.

Still, she's faced some adversity, even at age five. On more than one occasion, we've had to reassure her that girls are welcome as scientists, after somebody told her that science was for boys. Such unenlightened attitudes demand firm and immediate pushback, and I hope that if you know a young girl, you will encourage her to discover and make use of her inner scientist as well.

SCIENCE IS NOT JUST FOR THE RICH

Most news stories about modern science tend to talk about huge, expensive experiments—the $10 billion Large Hadron Collider, the $2.5 billion Curiosity rover on Mars, huge and expensive drug trials, and so on—giving the impression that science is inherently very expensive. On top of that, the large amount of scientific knowledge accumulated over the last few hundred years means that working on cutting-edge research requires a considerable length of time in school, which again can rack up a large price tag very quickly. And direct instruction in science can be very expensive to provide in schools, leading some school districts to deemphasize science as funding gets tight. These factors, among others, can combine to create an impression that a career in science is an opportunity only available to those who are already comfortably well off.

Again, this idea is false and has always been false. The British physicist Michael Faraday came from humble origins, in an age even more sharply divided by class than our own, but he became one of the greatest scientists of the nineteenth century.*
As a bookbinder's apprentice, Faraday had little formal education, but he taught himself a good deal by reading the books that came through the shop and attending public lectures by the scientists of the day. In an effort to escape a career binding the books of others, he assembled and bound a careful set of notes from a series of such lectures by Sir Humphrey Davy and delivered it to Davy. A short while later, Faraday was hired as a combination laboratory assistant and personal valet to Davy.†

From this humble start, Faraday built an amazing career in science. He distinguished himself as the best of Davy's assistants and was quickly promoted to a full-time scientific role, where his exceptional care and precision in carrying out experiments helped him make foundational discoveries in chemistry, optics, and electromagnetism. His greatest achievement is what is now known as Faraday's law, showing how changing magnetic fields generate electricity. This is the basis for nearly all modern electrical power generation, which uses rapidly rotating magnets driven by falling or boiling water to induce a voltage in coils of wire. Modern technological civilization as we know it would be impossible without Faraday's discovery. He also introduced the idea of electric and magnetic fields, a new way of looking at electromagnetic phenomena that eventually led to Maxwell's equations of electromagnetism.

---

* Albert Einstein famously kept three pictures of great physicists in his office: Isaac Newton, James Clerk Maxwell, and Faraday.
† This dual position led to some awkwardness, particularly on a tour of Europe, when Davy's new wife repeatedly snubbed Faraday, making him ride outside their carriage and eat with the servants.

Some might argue that things were simpler in Faraday's time and that the growth of modern science has made such a career arc improbable. Even well into the modern era, however, significant scientific developments continue to require few resources beyond a keen eye and a sense of curiosity. In 1963, Erasto Mpemba, a schoolboy in Tanzania, noticed that when he threw a batch of ice cream mix into the freezer straight after boiling it, his ice cream solidified before a classmate's room-temperature mixture that was put in at the same time. This seemed to run counter to what he was taught about the heating and cooling of objects, but when he asked his teacher about it, the boy's observation was mocked as "Mpemba's physics, not the universal physics."

He kept investigating the phenomenon he had observed, though—repeating the experiment in his school's freezer and finding that ice cream vendors in the area regularly used the same effect to speed production. Mpemba kept being teased about his investigations until he posed the question to Denis Osbourne, a visiting professor from a university in Dar es Salaam. Osbourne, like Mpemba's teachers, was skeptical, but he was intrigued by the boy's insistence that the effect was real and, when the professor returned to his lab, he asked a technician to check out the observation. The technician confirmed that hot water placed in a freezer had formed ice before a colder sample placed in at the same time, but the assistant promised to "keep on repeating the experiment until we get the right result."

In 1969, Osbourne and Mpemba coauthored a paper on what is now known as the Mpemba effect, which has been the subject of much lively discussion over the last forty-five years. Numerous experimental investigations have revealed a range of subtleties to the effect, and attempted explanations have included the effects of convection, dissolved gases, supercooling of the water, and, in November 2013, a paper claiming that the effect arises from fundamental properties of the structure of water. The Royal Society of Chemistry ran a contest for the best explanation of the Mpemba effect, with the £1000 prize awarded by Mpemba himself in 2013, but the question is obviously far from settled.

While expensive equipment and education make it easier to do some kinds of science, the essential activity is mental. All you really need to do science is careful observation and a systematic approach to thinking about the world—Mpemba's discovery didn't require highly complicated apparatus, but just needed a keen eye and an open mind. Those resources are available to anyone, regardless of economic status. Everybody, rich and poor alike, has an inner scientist, and if you listen to what people have to say about the world, you can make amazing discoveries.

SCIENCE IS NOT A CULT

One of the more popular forms of inveighing against science for ideological reasons is the declaration that scientists have set themselves up as a "modern priesthood" and that science is effectively just another religion. Scientists, according to this view, exploit the arcane nature of their subject to baffle the public (and to indoctrinate junior scientists into the "faith") with incomprehensible graphs and numbers to push an insidious agenda. Depending on the political persuasion of the speaker, the scientific agenda involves advancing the interests of either godless, tree-hugging communists or rapacious corporations. Either way, according to these naysayers, science is bad for those outside the "priesthood."

As I hope this book has made clear, this attitude misrepresents the essential nature of science. Science isn't a collection of facts to be accepted on faith; it's a process for testing and generating reliable information about the world. And that process is available to everyone—there's no essential hierarchy involved.

This doesn't mean you're required to carry out all your own experiments and simulations to duplicate a new scientific result—that would be wildly impractical. But even if you don't have access to all the tools and techniques needed to reproduce a result in detail, you should be able to follow the reasoning process involved. A new scientific result will compare observations with the predictions of a model and will either support or refute part of that model. When reading reports of new discoveries, you can trace out this argument and evaluate its components for yourself. Does the described technique seem reliable? Do the results fit with previous well-established observations? Does the model used to explain the observations allow useful predictions? If necessary, you can work backward and apply the same method to earlier reports whose results or techniques are needed to understand the latest work. By tracing the steps of the argument and using a bit of common sense, you can often make reasonable judgments about the overall reliability of a new result.

All these steps of logic may seem a daunting prospect, but we do something similar whenever we encounter anything new, even outside science. When I was in graduate school, the lab where I worked employed a lot of European postdocs, and our lab group would plan an outing to a minor-league baseball game to show the Europeans an essential bit of Americana. When we did this, we didn't expect them to read the rules of baseball from cover to cover before the first inning. Instead, before the game, we gave a basic outline of the rules so they could follow what was going on, and explained the difference between balls and strikes, fair and foul balls. Then, as play went on, we would address any complications that came up—a batter hit by a pitch, say, or a ball bouncing into the stands for a ground-rule double. By the fifth or six inning, they usually had a pretty good idea of what was going on and could make some reasonable judgments about the teams and players without needing to read all the rules.*

Learning new activities on the fly is something people do all the time, whether they are watching an unfamiliar sport for the first time, being introduced to a board game, or watching a new movie. We take for granted that people will be able to pick up the basics of a new interest and develop a more sophisticated understanding as complexities arise. Science is no different—the mass of accumulated knowledge underlying new scientific results may seem intimidating, but I can assure you that learning enough to follow an unfamiliar field of science is no worse than learning enough to keep up with my five-year-old's burgeoning interest in Pokémon.

Science is not something to be taken on faith, accepted without understanding. If you're treating scientific pronouncements as something to be believed without understanding, you're doing it wrong. And if the person trying to get you to accept a scientific idea isn't giving you the background necessary to evaluate the claim, *they're* doing it wrong. Accepting information on faith, or expecting others to do so, is antithetical to understanding a subject. One of the most important and regularly re-invented aphorisms in science is that you don't truly understand a subject unless you can explain it to a

---

* As long as the dreaded infield fly rule never came up, anyway. That one tends to confuse most Americans.

nonscientist.* As we've seen, the term *nonscientist* is a bit of a misnomer, as we all have inner scientists, but the general point is good: If you truly understand something, you should be able to lay out the essential facts and, more importantly, the process by which we know those, and help someone else reach the same understanding.

SCIENCE IS NOT FOR ROBOTS

Closely related to the idea of science as a religion is the common impression that belief in science must be all-consuming and displace everything else. In this view, scientists must be exaggerated versions of Vulcans from *Star Trek*, dedicated to logic above all else. Scientific knowledge and training supposedly renders scientists incapable of appreciating the beauty of art or experiencing normal human emotions. Sheldon Cooper of *The Big Bang Theory* is this sort of caricature, and his portrayal is one of the most controversial aspects of the show among physicists.

This is another unfortunate image that even a cursory knowledge of the history of science should dispel. While there are plenty of examples of socially awkward scientists—Paul Dirac and Henry Cavendish are the extreme examples here—most scientists have perfectly normal emotional lives. Most scientists also have interests outside science. Robert Oppenheimer, the director of the Manhattan Project, was notably well read, with a lifelong interest in philosophy and religion;

---

* Just within physics, variants of this line include the claim that you should be able to explain your research to a college freshman (attributed to Feynman), a barmaid (attributed to Rutherford), or your grandmother (attributed to Einstein).

he famously quoted the *Bhagavad Gita* after the detonation of the first atomic bomb. Richard Feynman dabbled in drawing, painting, and drumming. Even the famously awkward Dirac had outside interests, once surprising an interviewer by holding forth for several minutes about the newspaper comics *Prince Valiant* and *Blondie*.

Science is a process, a tool, something to be used as needed. It doesn't need to conflict with, or take the place of, other human activities or beliefs—even religious beliefs. Despite some forms of fundamentalist religious faith that are more or less impossible to reconcile with science—there's a mountain of evidence against a strict literal reading of the book of Genesis or any other creation story—plenty of scientists successfully combine personal religious faith with careers in science. Some of the very best scientists I know are regular churchgoers and see no inherent conflict between science and religion.

Science is undeniably effective, and if you start making more use of your inner scientist, you may find value in taking a more scientific approach to a lot of things. But there's no obligation to have science displace all other interests. If you can find a way to reconcile science with other beliefs and it works for you, that's all you need; there's no special guild in science to disbar you or revoke your license.

As for the question of appreciating beauty in art and nature, like a lot of scientists, I've always been somewhat puzzled by the idea that art and science are opposites. To be sure, there are some places where my knowledge of science undermines my ability to enjoy works of fiction, but then people who talk about scientists being unable to appreciate great art don't usually have scientifically inaccurate superhero movies in mind. Beyond that, though, there's no conflict between knowing something about the workings of the universe and appreciating its beauty. Richard Feynman put it well in a famous footnote to the *Feynman Lectures on Physics*:

> Poets say science takes away from the beauty of the stars—
> mere globs of gas atoms. I too can see the stars on a desert
> night, and feel them. But do I see less or more? The vast-
> ness of the heavens stretches my imagination—stuck on
> this carousel my little eye can catch one-million-year-old
> light. A vast pattern—of which I am a part. . . . What is the
> pattern, or the meaning, or the why? It does not do harm
> to the mystery to know a little about it. For far more mar-
> velous is the truth than any artists of the past imagined it.
> Why do the poets of the present not speak of it? What men
> are poets who can speak of Jupiter if he were a man, but if
> he is an immense spinning sphere of methane and ammo-
> nia must be silent?

There is also the final paragraph of Darwin's *On the Origin of Species*, rightly hailed as a poetic expression of the scientific worldview:

> It is interesting to contemplate an entangled bank, clothed
> with many plants of many kinds, with birds singing on the
> bushes, with various insects flitting about, and with worms
> crawling through the damp earth, and to reflect that these
> elaborately constructed forms, so different from each
> other, and dependent on each other in so complex a man-
> ner, have all been produced by laws acting around us. . . .
> [F]rom the war of nature, from famine and death, the
> most exalted object which we are capable of conceiving,
> namely, the production of the higher animals, directly fol-
> lows. There is grandeur in this view of life, with its several

powers, having been originally breathed into a few forms or into one; and that, whilst this planet has gone cycling on according to the fixed law of gravity, from so simple a beginning endless forms most beautiful and most wonderful have been, and are being, evolved.

A scientific understanding of the world need not displace a sense of wonder; in fact, understanding often enhances wonder. The vast sweep of the universe and the interrelationship of all living things are as awe-inspiring in their own way as anything in the history of art.

Science is not for robots, and your inner scientist is no enemy to human passion. Scientists are not required to renounce ordinary emotions—on the contrary, many scientists find a deep and emotional connection to their subject. There's perhaps no better illustration of this than the story from which this book takes its title: When Archimedes had the original eureka moment, legend has it he leaped from the bath, so excited he forgot to dress before running through the streets, naked and dripping, to tell the king about his discovery.

SCIENCE IS NEVER OVER

When physics gets coverage in the mass media, it's almost always focused on developments in particle physics, either experimental projects like the discovery of the Higgs boson at the LHC or speculative theoretical endeavors like string theory. Such a focus almost always casts the project of physics as a search for a "theory of everything," an elegant encapsulation of all the known particles and interactions in the universe. This description sometimes leads people to ask what will become of physics if physicists ever succeed. If they find the theory of everything, does physics as we know it come to an end?

I'm always kind of bemused by this question, because as I explained before, my research field is atomic, molecular, and optical (AMO) physics. The foundational moment for AMO physics happened just over a hundred years ago, with Bohr's initial quantum model of hydrogen, and the theoretical and mathematical apparatus of atomic physics was mostly complete by 1950. And yet, there are very nearly as many physicists studying AMO physics as are studying particle physics—the American Physical Society's Division of Atomic, Molecular and Optical Physics has just over three thousand members; the Division of Particles and Fields, just over thirty-five hundred. Both groups are handily outnumbered by condensed-matter physicists (who study the behavior of solids and liquids)—the Division of Condensed Matter Physics boasts just over six thousand members—another group whose essential theories were pretty well set in the middle of the last century.

And of course, you can extend this reasoning up the hierarchy of scientific complexity. Chemistry, after all, is just physics with larger numbers of atoms. Biology is all about the chemistry of molecules inside cells, and geology is just a combination of chemistry and physics. All of these sciences depend on interactions whose basic properties have been understood for the better part of a century, and all these fields are thriving, employing many more scientists than are involved in the pursuit of the theory of everything.

The mere fact that you can write down an equation describing the basic properties of an interaction does not mean you have exhausted its possibilities. If anything, as you bring more and more particles together to interact, the possibilities expand exponentially. To borrow the title of a famous 1972 article by Nobel laureate Phillip Anderson, "More Is Different," complex systems, even when their components interact by simple rules, provide a nearly infinite range of possible phenomena. Working out and observing all the possible interesting combinations of atoms and molecules will keep scientists happily occupied essentially forever. Even if all the string theorists were to pack it in after finding their theory of everything, it would hardly make a dent in the day-to-day practice of science. We're never going to run out of interesting phenomena to investigate; science will never be over.

There's another sense in which science is never over as well, which is that all scientific results are provisional. Many popular treatments of the history of science include some version of a statement, attributed to a late-nineteenth-century scientist, that nothing new remained to be discovered.* The quote is usually presented as an example of hubris—although scientists of the late 1800s had a good deal to be proud of, most of what they thought they knew, especially in physics, was about to be profoundly shaken by the dual revolutions of relativity and quantum mechanics. The statement can also serve as a cautionary tale, though, a reminder that everything we think we know is just a few experiments away from being turned completely around and cast in a new light. The core of the physics of the 1900s still survives and is taught, but we recognize now that it's only an approximation of a deeper theory and there are numerous phenomena it can't explain.

The same might be true of modern science. The models we have at present are too good at predicting reality to ever

---

* The most common specific attribution is to the great British physicist William Thomson, Lord Kelvin. There's no solid evidence that he ever actually said this, though.

be completely discarded—some calculations in QED are good to fifteen decimal places—but they might be only a useful but limited approximation of a deeper theory. And the same would be true of a theory of everything. No level of proof can ever truly lift a model beyond reproach. Any scientific model, even one that has been established for hundreds of years, is always one experimental test away from being completely overhauled.

I said at the start of this book that acknowledging your inner scientist is important because the scientific mind-set is an optimistic and empowering approach to the world. If you can think like a scientist—and I hope the preceding pages have convinced you that you can, and indeed, already do—then you never need to settle for ignorance. You can always use the process of science to find an answer.

This idea might represent the most important thing that science is not: Science is never over. Science is not a fixed track with a definite end point; it's an open-ended process that goes on forever. There will always be new questions to ask, new models to invent, new experiments to perform, and new results to share. No matter who you are or where you're from, if you're willing to embrace your inner scientist, you'll find the world an endless source of wonder and amazement.

SCIENCE IS NOT (JUST) A SPECTATOR SPORT

I hope that through the examples presented in this book, I've convinced you that even if you don't think of yourself as a scientist, you use scientific thinking every day. If you play sports, or cards, or hidden-object games, or work crossword puzzles, or even just watch mystery shows on television, you are making use of the mental processes that enabled some of the greatest discoveries in the history of science.

Now that you recognize the processes shared by both scientific discovery and recreational activities, you should be in a better position to understand and evaluate news stories about scientific discoveries. That's not to say that you will necessarily understand all the details, particularly in the more mathematical sciences. But I hope this book has better equipped you to recognize the *process* of the discovery.

As I write this in March 2014, the story dominating science news coverage is the announcement that the BICEP2 collaboration has discovered evidence of gravitational waves produced in the first trillionth of a trillionth of a trillionth of a second after the Big Bang. The science of the discovery is extremely complicated, and to be honest, I don't understand it all that well myself. But I can follow the story by mapping it to the process of science. The BICEP2 scientists *looked* out at the universe and noticed a pattern in the images from their telescope. They *thought* about possible effects that might explain the observed pattern, which included primordial gravitational waves. They *tested* their model by comparing their observations to both their preferred explanation and other effects that might produce similar patterns. And then they *told* the world about it, allowing other scientists to do further tests and allowing scientists and nonscientists alike to share in the wonder of this glimpse of the primordial universe.*

More than merely appreciating news stories about science, though, I hope that recognizing the many ways you use your

---

* Other scientists responding to the results follow a similar pattern: They look at the BICEP2 results, think about alternative explanations, test the alternatives to see if they hold up, and report both the alternative explanation and the results of the tests. The BICEP2 papers aren't the final word, but they are the start of a discussion that will almost certainly still be going on well after this book is published.

inner scientist in everyday life will empower you to make more conscious use of your inner scientist, to become an active participant in the scientific process, not just a spectator on the metaphorical sidelines. This might mean contributing to ongoing research through citizen-science projects like those discussed in Chapter 4, but you can also use science on a more personal and local scale.

So, *look* at the world around you, and if you notice things that don't seem right—injustices, inefficiencies, or just oddities—investigate them further. *Think* about why and how those things might happen and what you could do to change them for the better. *Test* your ideas by trying new things, making more observations, refining your models, and repeating the process. And when you find something that works to explain or improve your situation, *tell* everyone about it, so we can all benefit.

From the very dawn of humanity, from making stone tools and mixing pigments in Africa a hundred thousand years ago, to ancient astronomers stacking giant rocks to track the rising Sun, to the modern world of Angry Birds and astronomical observations back to an infinitesimal instant after the birth of the universe, science has always been an essential human activity. The efforts of countless generations of scientists have made us what we are today and will shape our future, for good or ill. There's room for every one of us to participate in this process, if we choose. I hope you'll acknowledge and employ your inner scientist and join in to make a better future for everyone.

ACKNOWLEDGMENTS

This book covers a much wider range of topics than my previous books, including some topics well outside my professional training. As a result, it has been a good deal more difficult to write. I am extremely grateful for the patient assistance of many colleagues, friends, and acquaintances who read and commented on draft chapters and answered my dumb questions about their fields of expertise. These include (but are not limited to) Michael Bradley, Chris Chabris, Thony Christie, Tomasso Dorigo, Nick Hadley, Becky Koopmann, Jon Marr, Rich Meisel, Trey Porto, Dave Pritchard, Simon Rainville, Josh Shapiro, Jason Slaunwhite, Tom Swanson, Brian Switek, and Scott Turner. This book is vastly better than it would have been without their generous assistance; any errors that might have crept in originated with me, not them.

Many of these ideas got a test drive of sorts on my blog, *Uncertain Principles*, where the comments of my readers and other bloggers helped shape some of the argument. Many thanks to them, and to the folks at ScienceBlogs and National Geographic for keeping the site running smoothly. Large chunks of this book were written at the Starbucks in Niskayuna, New York; many thanks to Angelina, Brian, Christine, Cody, Gina, Kurt, and Phil for keeping me supplied with caffeine and Wi-Fi. Thanks, too, to the staff at Shaffer Library at Union College; I promise I'll return at least some of your books one of these days.

This book went from a vague set of loosely connected blog posts to a coherent proposal thanks to some great suggestions from my agent, Erin Hosier. It has become an actual book thanks to my editor, T. J. Kelleher, who helped shape this into a much stronger argument, and the great folks at Basic, particularly Sandra Beris and Patricia Boyd, who make me look smarter than I would otherwise.

The writing of this book would have been nearly impossible without my wife, Kate Nepveu, who, among other things, listened patiently as I talked through rough ideas, read and commented on innumerable drafts, talked calmly during the occasional moments of sheer panic, and dealt with the kids when I had to scramble to meet deadlines. Thanks to Claire and David as well, for providing cheerful distractions at some key junctures and being surprisingly tolerant of Daddy sitting at the computer typing all the time.

But the biggest thanks are due to my parents, who over the years have provided everything from answers to my earliest questions to emergency babysitting services on next-to-no notice. They have always encouraged and supported my interest in science, wherever that led, even when they had no idea what the heck I was talking about. None of this would have been possible without them.

INDEX